LEADERSHIP PRACTICES FOR HEALTHY LANDS AND WATERS

Don Kent, PhD

Foreword by Andrew Bennett
Founder and President of the Bennett Performance Group

Praise for Leadership Practices for Healthy Lands and Waters

Don Kent's Leadership Practices for Healthy Lands and Waters is a comprehensive, well-researched guide on how to develop and maximize leadership skills to protect nature. Many conservation professionals, either willingly or reluctantly, have been promoted to leadership positions with little to no training. When training is offered, after the fact, it only covers limited aspects of the multiple dimensions of leadership in the real world. Leadership Practices for Healthy Lands and Waters offers the theoretical underpinnings of the topics covered, thorough insights into all aspects of leadership, and helpful real-world examples. Everything you need to become a better leader is here. The world needs more effective leaders, and resources like Leadership Practices for Healthy Lands and Waters give us hope for a better future for humans and nature. - *Ricardo Stanoss, Director - Center for Learning Innovation at Smithsonian's National Zoo and Conservation Biology Institute.*

So, what you have in Don Kent is an educated, experienced scientist; a well-practiced strategist; and a leadership development expert. What a combination! ... when I spend time with Don, I suddenly feel like we can change things so that our people and planet can thrive together. I believe it's possible because Don thinks it's possible. He gives me hope. And isn't that one of the most important things a leader can give us? - *Andrew Bennett, Founder and President, Bennett Performance Group.*

I worked with Don for several years and found him to be a thoughtful and insightful colleague. With his new book, Don has taken his insights in land conservation and protection and aligned them with leadership thought to craft a valuable reference for a better understanding of the value of engagement. *Shane Ford, Director of Strategic Directives for the Northeast Region, British Columbia, Canada.*

HEALTHY LANDS PUBLISHING

Library of Congress Cataloging-in-Publication Data

Kent, Don, 1955 –

Leadership Practices for Healthy Lands, 3rd Edition, by Don Kent, Ph.D.

Includes bibliographical references and index.

ISBN: 978-1-7377473-2-1

1. Leadership, 2. Conservation

DEDICATION

I thank my beloved wife, Karen Thompson, for her support and patience. Leadership is hardest with those we love the most.

Your purpose in life is to find your purpose and give your whole heart and soul to it.

– Gautama Buddha

TABLE OF CONTENTS

FORWARD

In 2014 I was invited to speak at NatureServe's annual leadership conference. I had never heard of them, but when I learned that they use science to guide environmental conservation, I was interested. Since 1992, I have worked with corporations like Google, Marriott, and Ford Motor Company helping leaders create workplace cultures where people can thrive. My life's work is to revitalize the human spirit, particularly in the workplace. The opportunity to speak to an organization like NatureServe, which works to protect and preserve our planet, was in a different, inspiring realm.

To help me prepare for my presentation, I was introduced to Don Kent. Fortunately, NatureServe is based near my home, making it possible to meet with Don in person over lunch. I was immediately struck by Don's humility and deep passion for people's well-being and our planet. Over the course of lunch and many conversations that have followed, I have grown to respect Don deeply.

Don's one of those people who seem to be born with a purpose. His work is not work; it's a life mission. I don't know how Don feels about this, but I could imagine him never retiring. From a very early age, Don knew he wanted to dedicate his life to making our world a better place. Academically, he's a scientist (biology/zoology), and his first jobs were hands-on applications of science to guide responsible and sustainable approaches to projects impacting the environment.

Over time, Don began to appreciate the impact of leadership and strategy to make truly meaningful change. His early academic discipline and voracious appetite for learning led him to study leadership and strategy. He began teaching leadership and eventually designed comprehensive leadership development programs. He also facilitated numerous strategic planning initiatives using thoughtful approaches designed to create engagement and ownership within organizations.

So, what you have in Don Kent is an educated, experienced scientist; a well-practiced strategist; and a leadership development expert. What a combination!

Leadership Practices for Healthy Lands and Waters delivers all of Don's wisdom and experience right to your hands. You will be better equipped as a steward of our planet when you read, reflect, and apply what Don shares. You will be a better leader, someone who inspires others to be their best so that you can accomplish more together. And you will be a better person because everything Don shares in this book is about being a better person.

It's appropriate that the final chapter of this book is entitled *Hope.* In my 27 years of speaking, teaching, and coaching about leadership, I have arrived at the same place Don has—that hope is essential to life, and it's one of the most important things a leader can cultivate in others.

The final chapter on hope is the perfect send-off because every encounter I have with Don leaves me feeling hopeful. Dealing with environmental sustainability is daunting to consider. I get overwhelmed even thinking about the magnitude of the task. But when I spend time with Don, I suddenly feel like we can change things so that our people and planet can thrive together. I believe it's possible because Don thinks it's possible. He gives me hope. And isn't that one of the most important things a leader can give us?

Andrew Bennett, Founder and President
Bennett Performance Group

PART 1
THE CASE FOR CHANGE

THE OVERRIDING TENSION

The health of ecosystems on which we and all other species depend is deteriorating more rapidly than ever. We are eroding the very foundations of our economies, livelihoods, food security, health, and quality of life worldwide.

Intergovernmental Science-Policy Platform on Biodiversity and Ecosystem Services: 2019 Global Assessment Report on Biodiversity and Ecosystem Services

The Value of Our Lands and Waters

A tension exists between our need for development and healthy lands and waters. South Florida's agriculture and urbanization compete with fish and wildlife habitat, recreation, tourism, drinking water, and cultural heritage. Mekong River watershed deforestation, pollution, and hydropower diminish fisheries, forest resources, and the transport of goods and services. Cattle ranching and channelization in South America's Pantanal reduce flood control, fisheries, wildlife, and tourism. Increasing greenhouse gases threaten every good and service provided by healthy lands and waters and in turn economies and human health.

Re-examining trade-offs to reduce the tension between developed and healthy lands and waters is essential. Without a doubt, developed lands are essential to our well-being and prosperity. Cities and towns, roadways, rail lines, dams, energy facilities, farms, ranches, and aquaculture operations meet our base needs for food, shelter, and safety, our psychological needs for belongingness, love, and self-esteem, and our need for self-fulfillment.[1] Developed lands are a critical element of our global economy, generating about US$88 trillion per year.[2]

But healthy lands and waters also constitute an efficient and effective solution for meeting our base, psychological, and self-fulfillment needs. Monetarily, healthy lands and waters are valued at US$170–$190 trillion per year, more than twice the value of developed lands and waters.[3] Healthy lands and waters provide essential goods and services and meet our social, cultural, spiritual, and religious needs.[4,5]

Ecosystem Services

The value of healthy lands and waters is related to their inherent functions. Meadows, wetlands, rivers, lakes and ponds, deserts, mountains, tundra, coastlines, and seas have hydrological, geological, biological, and chemical attributes. These attributes combine to optimize plant productivity, soil fertility, water quality, and atmospheric chemistry. In turn, providing a habitat for wildlife, fish, and other aquatic vertebrates and invertebrates, recharging groundwater, cycling nutrients, and altering and storing floodwaters.[6,7]

The term used to describe the direct and indirect contributions of healthy lands and waters to human well-being is *ecosystem services.*[8] Ecosystem services are conventionally categorized as regulating, cultural, habitat/supporting, and provisioning.

1 A.H. Maslow. 1943. A theory of human motivation. Psychological Review (50):370–396.

2 World Bank. Accessed 16 October 2002. https://data.worldbank.org/indicator/NY.GDP.MKTP.CD.

3 Naturschutzbund Deutschland, Boston Consulting Group, and Birdlife International. 2020. The Biodiversity Imperative for Business: Preserving the Foundations of Our Well-Being. https://www.nabu.de/imperia/md/content/nabude/biodiv/200923-nabu-bcg-studie-biodiv-englisch.pdf.

4 Intergovernmental Science-Policy Platform on Biodiversity and Ecosystem Services. 2018. Nature's Contributions to People (NCP) – Article by IPBES Experts in Science <https://www.ipbes.net/news> (Intergovernmental Platform on Biodiversity and Ecosystem Services (IPBES), Bonn, Germany, 2018.

5 Sandra Díaz, Unai Pascual, et al. 2019. Assessing nature's contributions to people. Science. August. Vol. 359, Issue 6373, pp. 270–273.

6 David Pearce and Dominic Moran. 1994. The Economic Value of Biodiversity. Earthscan Publications Limited, London.

7 Robert Costanza, John Cumberland, Herman Daly, Robert Goodland, and Richard Norgaard. 1997. An Introduction to Ecological Economics. 1st Edition. CRC Press, Boca Raton, Florida.

8 Millennium Ecosystem Assessment, 2005. Ecosystems and Human Well-being: Synthesis. Island Press, Washington, DC.

Regulating services moderate natural phenomena and account for 65 percent of ecosystem value. Healthy lands and waters regulate climate, purify water, decompose waste, prevent soil erosion, moderate floods, and pollinate plants. These are services that are not always obvious, but which cannot be easily or inexpensively replaced.

Cultural services provide 18 percent of ecosystem value and consist of nonmaterial contributions to the development and cultural advancement of people. For example, the role of healthy lands and waters in local, national, and global cultures, and how knowledge builds and is shared. Cultural services also include recreation and tourism.

Ten percent of ecosystem value is from habitat and supporting services. These include photosynthesis, water and nutrient cycling, and soil creation services that allow the planet to sustain life and enable regulating, cultural, and provisional services. Habitat or supporting services also include refugia and nurseries for animals and plants, and maintenance of plant and animal gene pools.

Provisioning services account for 7 percent of ecosystem value. They include consumable goods like food from land and sea, genetic material for domesticating and maintaining plants and animals, medical resources derived from plants and animals, and raw materials like forest products, natural gas, and oils.

Regardless of the ecosystem service category, we value healthy lands and waters we use now and those we do not currently use but may want to use later. We value lands and waters we may never use because they might benefit someone else in the future. And we value lands and waters simply because we are comforted knowing that an attractive land or a charismatic animal exists.

Healthy lands and waters have physical, social, and emotional value.[9,10] They reduce our hospital stays and the number of painkilling drugs we need. Healthy lands and waters help us recover from mental fatigue and

9 Rachel and Stephen Kaplan. 1989. The Experience of Nature: A Psychological Perspective. Cambridge University Press, Cambridge, England.

10 Richard Louv. 2011. The Nature Principle. Algonquin Books of Chapel Hill, North Carolina.

psychophysiological stress. They make us feel better by slowing our heartbeat, lowering our blood pressure, and promoting relaxed brain wave patterns. Healthy lands and waters increase self-esteem, happiness, educational status, and emotional development in at-risk children and others.

Healthy lands and waters even make our homes more comfortable. The rustle of leaves and the wind blowing through branches create a pleasing background sound and reduce noise. Homes in healthy lands are cooler in the daytime and warmer in the nighttime. And when it comes time to sell, your property values are 5 to 20 percent higher than homes in less ideal settings.

The Decline of Healthy Lands and Waters

The values of both developed and healthy lands and waters argue for a circumspect, balanced plan for global and local development and conservation of natural resources. Yet, despite our dependence on healthy lands and waters their destruction, degradation, and unsustainable use continue at an unprecedented rate.

Humans have changed ecosystems more rapidly and extensively over the last 50 years than in any comparable period in our existence, mostly in response to our need for food, fresh water, timber, fiber, and fuel.[11] Ten to 20 percent of the planet's land is degraded. Fourteen of 24 ecosystem services examined by the Millennium Assessment, a comprehensive examination of ecosystems and human well-being, are degraded or degrading.

One million animal and plant species are threatened with extinction, including those domesticated for food and agriculture.[12] Ninety percent of the world's seabirds have fragments of plastic in their stomach.[13]

11 Millennium Ecosystem Assessment. 2005. Ecosystems and Human Well-being: Synthesis. Island Press, Washington, DC.

12 Intergovernmental Science-Policy Platform on Biodiversity and Ecosystem Services. 2019. Global Assessment Report on Biodiversity and Ecosystem Services.

13 World Wildlife Fund. 2018. Living Planet Report – 2018: Aiming Higher. Grooten, M. and Almond, R.E.A.(Eds). WWF, Gland, Switzerland.

The declining health of our lands and waters impacts nearly every dimension of human health and well-being.[14]

- Our rainwater and one-third of our tap water contains contain forever chemicals – human-made chemicals that don't occur in nature.[15,16]
- Each year, one in every four deaths, more than 12 million people, are due to air, water, and soil pollution, chemical exposure, climate change, and ultraviolet light.[17]
- Of 7.8 billion people in the world, 750 million lack adequate access to clean drinking water, and 805 million do not have enough food to eat.[18]
- Around 9 million people die of hunger and hunger-related diseases every year, many due to climate change-induced drought.[19]
- In the Mekong River watershed dams are affecting the health and livelihood of 60 million people.[20]
- Excessive nutrient loading diminishes water quality, leading to drinking water shortages, coastal dead zones, and collapsing fisheries.
- Climate change and habitat loss is reducing biological diversity at a rate 100 to 1,000 times higher than the naturally occurring

14 Samuel Myers, et al. 2013. Human health impacts of ecosystem alteration. PNAS. www.pnas.org/cgi/doi/10.1073/pnas.1218656110.

15 Ashira Morris. 2021. EPA to regulate and restrict toxic "Forever" PFAS chemicals. Our Daily Planet. https://www.ourdailyplanet.com/story/epa-to-regulate-and-restrict-toxic-forever-pfas-chemicals/?mc_cid=06c4f234dd&mc_eid=86708a74a7.

16 Rosie Frost. 2022. Rainwater everywhere on Earth unsafe to drink.due to "Forever Chemicals" study finds. Euronews. https://www.euronews.com/green/2022/08/04/rainwater-everywhere-on-earth-unsafe-to-drink-due-to-forever-chemicals-study-finds

17 World Health Organization. 2016. An estimated 12.6 million deaths each year are attributable to unhealthy environments. https://www.who.int/news-room/detail/15-03-2016-an-estimated-12-6-million-deaths-each-year-are-attributable-to-unhealthy-environments.

18 Do Something.Org. 2019. 3 September https://www.dosomething.org/us/facts/11-facts-about-global-poverty.

19 Mercy Corps. No date. Quick facts: What you need to know about global hunger. https://www.mercycorps.org/articles/quick-facts-what-you-need-know-about-global-hunger.

20 Hannah Beech. 2020. Below a new dam, a grim turn for the Mekong. 16 February. *New York Times*. https://www.nytimes.com/2020/02/15/world/asia/mekong-river-dams-thailand.html.

background extinction rate, a rate that is expected to accelerate in the coming years.[21,22]

- We are losing ecosystem services, the value provided by healthy lands and waters, at the rate of US$4.3–$20.2 trillion per year.[23]

A Path Forward

Our present course looks bleak for future generations, but all is not lost. Healthy lands and waters that provide ecosystem services by retaining their inherent hydrological, geological, biological, and chemical attributes can be sustained and restored, meeting our demand for the services they provide.

The Intergovernmental Science-Policy Platform on Biodiversity and Ecosystem Services calls for transformative change, a fundamental, system-wide reorganization across paradigms, goals, and values.[24] Noted environmentalist Erle Ellis posits that a better future for nature and people is about motivating people to work together and creating the social processes, institutions, and strategies to make planetary cooperation possible.[25]

The Millennium Assessment evaluated scenarios for just such an outcome, concluding that a globally connected society and actions focused on regional watershed-scale ecosystems offer potential. Less promising are scenarios relying on technology and highly managed ecosystems. A scenario envisioning a regionalized and fragmented world is unlikely to sustain healthy lands and waters and their ecosystem services.

21 Gerardo Ceballos, Paul R. Ehrlich, Anthony D. Barnosky3, Andrés García, Robert M. Pringle and Todd M. Palmer. 2015. Accelerated modern human-induced species losses: Entering the sixth mass extinction. Science Advances: 1(5) e1400253.

22 Jurriaan M. De Vos, Lucas N. Joppa, John L. Gittleman, Patrick R. Stephens and Stuart L. Pimm. 2015. Estimating the normal background rate of species extinction. Conservation Biology 29(2) 452–462.

23 Robert Costanza, Rudolf de Groot, Paul Sutton, Sander van der Ploeg, Sharolyn J. Anderson, Ida Kubiszewski, Stephen Farber, R. Kerry Turner. 2014. Changes in the global value of ecosystem services. Global Environmental Change 26:152–158.

24 Intergovernmental Science-Policy Platform on Biodiversity and Ecosystem Services. 2019. Global Assessment Report on Biodiversity and Ecosystem Services.

25 Erle C. Ellis. 2019. Sharing the land between nature and people. Science Magazine Digital. June.

Actions

What actions can we take to achieve transformative change, motivate people to work together, and support promising scenarios? Historically, we have relied on three fundamental practices – regulation, protection, and education. These practices have produced notable successes, including recovery of protected species, conservation of critical lands, and increased scientific literacy. Each will continue to be important moving forward.

But as illustrated by the continued decline of our lands and waters, regulation, protection, and education alone as currently practiced cannot guarantee the comprehensive healthy lands and waters needed to ensure well-being and prosperity.[26]

Why not? Regulation requires adequate enforcement and can be weakened or eliminated by shifts in political ideology. Protected lands and waters require significant capital outlay and ongoing maintenance, rarely encompass entire ecosystems, and are susceptible to activities beyond their borders. Education is place, age, and culture specific, and in the United States and many other countries, education is focused on global competitiveness.[27] Regulation, protection, and education are all ineffective when contrary activities are motivated by greed or base needs for food, water, warmth, rest, security, and safety.

So how do we build on past practices to avoid further declines and reverse historic destruction, degradation, and unsustainable use of our lands and waters? Understanding context is critical.

- Sustaining healthy lands and waters is complex, requiring science, policy, implementation, and adaptive management in response to dynamic stressors.

26 Rachel E. Golden Kroner. 2019. The uncertain future of protected lands and waters. Science 364:881–886.

27 U.S. Department of Education. 2019. https://www2.ed.gov/about/overview/mission/mission.html.

- We face a knowledge burden requiring lengthy education or narrowing expertise, both of which reduce individual innovative capacity.
- We have insufficient quality data about threats to healthy lands and waters.

Success demands that we develop the expertise and skills to meet these challenges, and to test and implement promising scenarios and augment existing regulation, protection, and education practices. Three areas seem promising.

Collaborate and Innovate

First, we can adapt our formal and individual education to de-emphasize global competitiveness and favor a goal that prepares students to tackle large-scale challenges through collaboration, creativity, and innovation. A goal enabled by problem-based social and emotional learning. A goal that produces solutionaries - people who develop practical, effective, and visionary solutions that are regenerative, healthy, and just for all.[28]

Heurism

Second, we can emphasize heurism instead of knowledge. For too long our education system has been designed to produce knowledge workers taught to follow a set of established instructions down a single pathway to one conclusion. These left brain leaning knowledge workers score well on IQ tests. But left-brained aptitude only accounts for 4 to 10 percent of career success.[29]

The learner of the future will have social, emotional, and technological skills.[30] We must provide people with the capacity to detect patterns and opportunities, to synthesize information, craft satisfying narratives, and to combine seemingly unrelated ideas into something new.

In other words, we must transition from an economy and a society built on the logical, linear, computer-like capabilities of the Information Age to an

28 Zoe Weil. 2016. *The World Becomes What We Teach*. Lantern Books. Brooklyn, New York.

29 Daniel Pink. 2006. *A Whole New Mind*. Riverhead Books (Penguin Group). New York.

30 Jacques Bughin et al. 2018. Skill shift: Automation and the future of the workforce. May. McKinsey Global Institute. McKinsey & Company.

economy and a society built on the inventive, empathic, big-picture capabilities of the Conceptual Age.[31]

Boundary Crossing

Third, we need to become boundary individuals and boundary organizations. Boundary was originally conceived to describe an organization that mediates between science and policy development.[32] More recent conceptualization envisions organizations that operate in multiple fields, on different scales, and with different stakeholders.[33] I maintain that the concept applies equally to individuals.

Within the context of healthy lands and waters, boundary individuals and organizations facilitate collaboration and information flow across the natural and social science disciplines, across science and technology, and ultimately across science, technology, and communities. Not only communities as social groups that reside in a specific locale and share governance, but also groups that share common characteristics or interests such as scientific, political, cultural, and religious communities.

Leadership

The shift from unimaginative to innovative, competitive to collaborative, knowledge to heurism, from logical and linear to inventive, empathic, and big picture, from specialization to boundary crossing requires skills associated with leadership.

- Skills that strengthen character, clarify values and beliefs, further decision-making, and embrace change.
- Skills that build trust, promote change, create a shared vision, and motivate and enable others.

31 Daniel Pink. 2006. *A Whole New Mind*. Riverhead Books (Penguin Group). New York.

32 David Guston. 1999. Stabilizing the boundary between US politics and science: the role of the office of technology transfer as a boundary organization. *Soc. Stud. Sci. 29*(1), 87–111.

33 Karin Gustaffson and Rolf Lidskog. 2018. Boundary organizations and environmental governance: Performance, institutional design, and conceptual development. *Climate Risk Management* 19:1–11.

- Skills that facilitate effective communication, de-escalate stress, manage group dynamics, and identify and analyze stakeholders.
- Skills that promote creativity and innovation.
- Skills that build communities of practice.

LEADERSHIP

If your actions inspire others to dream more, learn more, do more and become more, you are a leader.

John Quincy Adams, 6th president of the United States

Leadership Practices Make a Difference

I had the good fortune at NatureServe, an international biodiversity data and science nonprofit, to develop and implement a leadership program for the more than 80 independent government, nonprofit, and academic organizations that comprise a Network. The curriculum was designed in collaboration with NatureServe and the Network to expressly meet individual, organizational, and Network needs.

The results were gratifying. At an individual level, alumni reported greater confidence, improved relationships, and better decision-making. Organizationally, the alumni were managing staff more effectively, had become more strategic, and were doing a better job demonstrating value to stakeholders. From a Network perspective, many of the alumni were contributing innovations, mentoring other members, participating in working groups, and serving on the NatureServe Board of Directors.

My experience reflects the results of study after study revealing the relationship between leadership practices and individual and organizational performance. Kouzes and Posner analyzed over 3 million responses to find that five practices – model the way, inspire a shared vision, challenge the process, enable others to act, and encourage the heart – make a difference in employee and organizational performance. Leaders who use the five practices: [34]

- Create higher-performing teams.
- Foster renewed loyalty and greater organizational commitment.
- Enhance motivation and willingness to work hard.
- More successfully represent their units to upper management
- Increase fundraising results.
- Extend the range of the organization's services.
- Increase retention and reduce absenteeism and turnover.
- Positively influence recruitment rates.

In a sample of 95 large companies with yearly revenues greater than $350 million and 10 years of financial performance data, organizations using the Kouzes and Posner five practices had net income growth 18 times higher, and stock price growth nearly three times higher, than organizations that did not use the practices.[35]

Of 1,435 good companies examined by Jim Collins over 40 years, 11 companies became great, outperforming the market by 6.9 times from 1996 to 2000.[36] The differentiator was level five leaders, those possessing a paradoxical combination of personal humility and professional will.

Harter et al. examined the relationship between employee engagement and organizational performance across 339 research studies, 230 organizations, 49 industries, 73 countries, and 1,882,131 employees.[37] Employee engagement had a substantial positive impact on profitability, productivity, employee

34 Jim Kouzes and Barry Posner. 2012. The Leadership Challenge. Jossey-Bass (Wiley). San Francisco, California. 25th Anniversary Edition.

35 Richard Roi. 2006. Leadership Practices, Corporate Culture and Company Financial Performance: 2005 Study Results. Crawford & Associates International. Palo Alto, California, US.

36 Jim Collins. 2001. Good to Great: Why Some Companies Make the Leap and Others Don't. Harper Business, New York, New York.

37 James Harter, Frank Schmidt, Sangeeta Agrawal, Stephanie Plowman and Anthony Blue. 2016. The Relationship Between Engagement at Work and Organizational Outcomes. Q12® Meta-Analysis: Ninth Edition. Gallup, Washington, D.C.

turnover, safety incidents, organizational downsizing, absenteeism, and product/service quality. Additional research is finding links between employee engagement and health and well-being.

Workplace engagement and commitment are significantly explained by how the leader behaves, and not at all by any characteristic of the constituents.[38] A leader's actions contribute more to such factors as commitment, loyalty, motivation, pride, and productivity than who the constituents are (i.e., age, gender, ethnicity, education), position, job, discipline, industry, nationality or, country of origin.

Leaders most clear about their values are more positive toward the workplace and rate their leadership effectiveness 25 percent higher than leaders less clear about their leadership philosophy.[39] As importantly, constituents rate leaders clear about their leadership philosophy 40 percent more effective than leaders who are not clear.

Managers who rate high in proactivity are assessed by their immediate supervisors as more effective leaders than managers rated lower in proactivity.[40] Proactivity consistently produces better results than reactivity or inactivity.[41]

Leaders encourage initiative in others. People experience 25 to 35 percent stronger feelings of pride, motivation, and team spirit when they are challenged and observe their leaders challenging themselves. Leaders who seek out challenges for themselves and constituents are evaluated 40 percent higher by constituents.[42]

Trust is a strong, significant predictor of employee satisfaction and productivity. Organizations with leaders who create a climate of trust

38 Jim Kouzes and Barry Posner. 2010. The Truth About Leadership. Jossey-Bass (Wiley). San Francisco, California. 197 pp.

39 Jim Kouzes and Barry Posner. 2012. The Leadership Challenge. Jossey-Bass (Wiley). San Francisco, California. 25th Anniversary Edition.

40 J. Michael Crant, J.M. and Thomas Bateman. 2000. Charismatic Leadership Viewed from Above: The Impact of Proactive Personality. Journal of Organizational Behavior 21(1):63–75.

41 Crant, J.M. 1995. The Proactive Personality Scale and Objective Job Performance Among Real Estate Agents. Journal of Applied Psychology 80(4): 532–537.

42 Jim Kouzes and Barry Posner. 2012. The Leadership Challenge. Jossey-Bass (Wiley). San Francisco, California. 25th Anniversary Edition.

outperform low-trust organizations by 286 percent.[43] Trust is the number one differentiator between the top 20 and bottom 20 percent of companies evaluated for corporate innovation.[44]

The real reason for organizational failure is typically not processes, performance management systems, or organizational structure, but employee behavior – the ability to get people to hold one another accountable to the process. In the best organizations, everyone holds everyone accountable through face-to-face conversations, regardless of position.

These findings, while largely focused on organizational outcomes, reflect human behavior and skills that can reasonably be extrapolated to working with stakeholders to achieve healthy lands and waters. The beneficial practices producing success are likely of greatest importance in situations in which the authority of one over others is absent.

What is Leadership?

The prosperity of the 20th century was due, in large part to an abundance of knowledge workers. Lawyers, doctors, accountants, engineers, and executives were trained to follow a set of established instructions down a single pathway to one conclusion. The emphasis was on supervision, organization, and group performance. Leadership was typically transactional in that the leader promoted compliance among their followers through reward and punishment.

A change in leadership began late in the 20th century, becoming visible to an expanded audience early in the 21st century with the publication of *Good to Great* by Jim Collins.[45] Collins's work illustrated a shift away from leadership as authority to leadership as responsibility. From leadership as an innate quality to leadership as an identifiable set of skills and abilities available to

43 Pamela Shockley-Zalabak, Sherwyn Morreale and Michael Hackman. 2010. Building the High-trust Organization: Strategies for Supporting Five Key Dimensions of Trust. Jossey-Bass. San Francisco, California, US.

44 PricewaterhouseCoopers. 1999. Innovation Survey. London.

45 Jim Collins. 2001. Good to Great: Why Some Companies Make the Leap and Others Don't. Harper Business, New York, New York.

anyone. For the first time, leadership was widely recognized as a reciprocal relationship between those who aspire to lead and those who choose to follow.

There is no universal definition of leadership. Fundamentally, a leader is someone who has followers, influences others, or guides individuals, teams, or organizations. Dwight D. Eisenhower, 34th president of the United States, said *leadership is the art of getting someone else to do something you want done because they want to do it.*

Well-intentioned, these definitions leave space for leaders with selfish interests. Raising expectations, Warren Bennis, founding chairman of the Leadership Institute at the University of Southern California, said *leaders are people who do the right thing as opposed to managers who do things right. Integrity became part of the equation.*

Vision is an essential leadership attribute. Warren Bennis describes leadership as the capacity to translate vision into reality. Mindy Gibbins-Klein, the award-winning entrepreneur and managing director of two profitable companies, said *leaders have a vision, share that vision, and inspire others to support that vision while creating their own.*[46]

Navigating change is another leadership essential. Kouzes and Posner define leaders as *men and women who guide others through adversity, uncertainty, hardship, disruption, transformation, transition, recovery, new beginnings, and other significant challenges.*[47]

Heifetz, Grashow, and Linsky distinguish between technical and adaptive leadership.[48] Technical leadership has known solutions that can be implemented using current knowledge, authority, and resources. Conversely, adaptive leadership applies to challenges that can only be resolved by exceeding current authority and changing people's priorities, beliefs, and habits.

46 http://www.mindygk.com/.

47 Jim Kouzes and Barry Posner. 2012. The Leadership Challenge. Jossey-Bass (Wiley). San Francisco, California. 25th Anniversary Edition.

48 Ronald Heifetz, Alexander Grashow and Marty Linsky. 2009. The Practice of Adaptive Leadership. Harvard Business Press. Cambridge, Massachusetts.

Leadership focused on change is sometimes referred to as transformational. Identifying a needed change, creating an inspiring vision, and implementing a valuable and positive change while working with others.

Leadership for Healthy Lands and Waters

Leadership for healthy lands and waters requires followers, integrity, vision, and change. My working definition of a leader for healthy lands and waters is ...

... someone who possesses the clarity of purpose, integrity, and competence to create trust and enable others to sustain lands and waters that promote well-being and prosperity for all.

The remainder of this book is organized to develop and implement leadership practices that support healthy lands and waters. The practices apply to other professions, avocations, and personal relationships.

Part 2, Prepare to Lead includes discovering your purpose, making better decisions, and embracing change.

Part 3, Leading Others includes leadership types, building trust, promoting change, creating a shared vision, and motivating and enable others.

Part 4, Share Leadership includes why you share leadership, communicating effectively, de-escalating stressful situations, managing group dynamics, and identifying and analyzing stakeholders.

Part 5, Create and Innovate includes the creative process, promoting creativity, and turning ideas into innovations.

Part 6, Build Communities of Practice includes peer consults, partner meetings, hosting communities, engaging citizen scientists, and working in socioecological systems.

Part 7, Hope for Healthy Lands and Waters concludes with why we should be hopeful and a plea that everyone should lead.

PART 2
PREPARE TO LEAD

DISCOVER YOUR PURPOSE

Your purpose in life is to find your purpose and give your whole heart and soul to it.

Gautama Buddha

Self-development is the essential base for your leadership, and authenticity is your cornerstone. Authenticity is self-awareness and a willingness to be transparent and known. Authenticity gives you the confidence to express ideas, choose direction, make decisions, and promote change. Authenticity enables the voice that represents who you are and what matters to you.

Authenticity requires clarity of purpose. The reason you do something and the force driving your determination and resoluteness. Many understandably stumble when asked to explain their purpose. We view daily life as immediate and concrete, and purpose as deferred and conceptual. The linking of your purpose with your daily life is how you find satisfaction. And satisfaction is necessary for the authenticity needed to lead.

You can achieve clarity of purpose in four steps. You begin by examining what characteristics you admire in others. This helps you reflect on your values, those ideas or behaviors to which you assign importance.

By understanding your values, you can identify your beliefs. These are the opinions or convictions that give you confidence in the truth or existence of something not immediately susceptible to rigorous proof.

Armed with your beliefs, and prodded by some leading questions, you can define your purpose. This is your mission in life that satisfies you, motivates you, and makes you authentic.

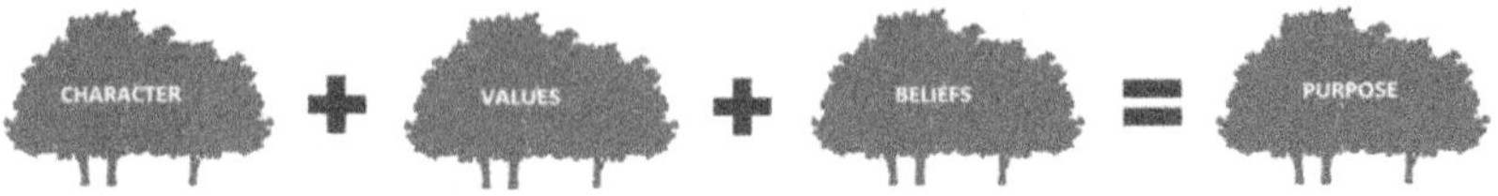

Characteristics

You learn from role models, people whose behavior, example, or success you emulate. Typically, your role models are people you know well. Initially, this may be a family member, and as you begin your career an immediate supervisor.[49] The latter has the greatest influence over your commitment to an organization, your ability to do your job, your ethics, and how you treat others.

What characteristics are you looking for in your role models?[50] In a study by Kouzes and Posner of 100,000 people around the globe, respondents were asked which characteristics they looked for in someone whose direction they would willingly follow.[51] Interestingly, the preferred characteristics were the same regardless of country, culture, ethnicity, organizational function, hierarchy, gender, level of education, or age group.

Four characteristics stood out – honest, forward-looking, competent, and inspiring.

Honesty is truthfulness in word and action. Forward-looking is having an eye for the future, rather than nostalgia or a focus only on the present.

49 Jim Kouzes and Barry Posner. 2012. The Leadership Challenge. Jossey-Bass (Wiley). San Francisco, California. 25th Anniversary Edition.

50 Milton Rokeach. 1973. The Nature of Human Values. Free press. New York, New York.

51 Jim Kouzes and Barry Posner. 2012. The Leadership Challenge. Jossey-Bass (Wiley). San Francisco, California. 25th Anniversary Edition.

Competent is the successful execution of your responsibilities. Inspiring is the ability to excite others to model your behavior or to follow a vision.

Quality/ Characteristic	Percent Response
Honest	87
Forward-looking	70
Competent	67
Inspiring	66
Intelligent	45
Fair-minded	41
Broad-minded	38
Supportive	36
Straightforward	34
Dependable	33
Cooperative	27
Courageous	25
Imaginative	24
Caring	22
Determined	22
Ambitious	18
Loyal	15
Mature	15
Self-controlled	9
Independent	6

The findings reflect the conclusions leadership pioneer Jim Collins[52] reached earlier when assigning characteristics to leaders. His lowest leadership level reflects minimal competence. The second and third levels reflect an increased ability to work with others. The fourth level demands a commitment to and pursuit of a compelling vision. The most accomplished leaders, level five, are distinguished by persistence balanced with humility.

52 Jim Collins. 2001. *Good to Great: Why Some Companies Make the Leap and Others Don't*. Harper Business, New York, New York.

If you do not feel yourself being increasingly humbled, you are not becoming a leader. Leadership is bestowed on us by others. A privilege and a responsibility, not an indication of our greatness.

Personal Values

What characteristics do you want others to see in you? Your characteristics reflect your values. These are the ideas and behaviors you deem important. I place a high value on integrity, so I hope that people characterize me as someone they can trust.

Values are your guides and when deeply held allow you to focus and make choices when decisions are difficult. The clearer you are about your values, the easier it is for you and those following you to commit to a chosen path. Leaders with clear values are more positive toward the workplace, rate their leadership effectiveness 25 percent higher, and are rated 40 percent more effective by their constituents.[53]

53 Jim Kouzes and Barry Posner. 2012. *The Leadership Challenge*. Jossey-Bass (Wiley). San Francisco, California. 25th Anniversary Edition.

To discover your values, rank the values listed below from 1 to 18, using 1 for the value most important to you and 18 for the value least important to you. Take your time and think carefully. The result should accurately represent how you feel about what is important in your life.

Value	**Rank**
Ambitious (hardworking and aspiring)	________
Broad-minded (open-minded)	________
Capable (competent; effective)	________
Clean (neat)	________
Courageous (standing up for your beliefs)	________
Forgiving (willing to pardon others)	________
Helpful (working for the welfare of others)	________
Honest (sincere and truthful)	________
Imaginative (daring and creative)	________
Independent (self-reliant; self-sufficient)	________
Intellectual (intelligent and reflective)	________
Logical (consistent; rational)	________
Loving (affectionate and tender)	________
Loyal (faithful to friends or the group)	________
Obedient (dutiful; respectful)	________
Polite (courteous and well-mannered)	________
Responsible (dependable)	________
Self-controlled (restrained; self-disciplined)	________

Which values are most and least important to you? Do they align with how you want to conduct yourself and the feedback you receive from others? Do your values align with and give rise to the honesty, forward-looking, competent, and inspiring characteristics sought by individuals willing to follow a leader?

Your value choices reflect your experiences and interpretation. There is nothing inherently wrong with any of the values, but your choices will define you.

The most important lesson of this exercise is for you to understand what is most important to you, and how it affects your relationship with others. A second lesson is that everyone may prioritize values differently. Understanding and accepting these differences will improve your communication, leadership, and success.

Honesty

Take a closer look at two values essential for leadership: honest and capable (aka competent). Both are essential if you are to trust yourself, be trusted, and persuade others to follow.

Honesty requires adherence to moral and ethical principles. Do you act with honesty? Take this self-evaluation.

Yes	**No**	
______	______	Am I honest in all my interactions with others?
______	______	Does my behavior match my intent?
______	______	Are my actions congruent with my values?
______	______	Am I more concerned with what is right than being right?
______	______	Do I act on good ideas even if they are not my ideas?
______	______	Do I embrace the truth rather than defend outdated positions?
______	______	Do I have the courage to do the right thing?
______	______	Do I make and keep commitments to myself?
______	______	Do I keep my promises?
______	______	Do I admit my mistakes?

______	______	Do I apologize when my mistakes harm someone?
______	______	Do I seek and accept feedback?

Do you have the honesty to lead? In practice, honest individuals are more concerned about what is right than being right, acting on good ideas than having good ideas, embracing the truth rather than defending outdated positions, building the team than personal aggrandizement, and recognizing contributions rather than being recognized.[54,55] Honest individuals seek mutual benefit, acting in the best interest of others to build trust.

Capable

Acting with honesty is not enough by itself. In a culture of shared responsibility, others will expect you to contribute. How capable are you? Take this self-evaluation.

Yes	No	
______	______	Does the work I am doing match my talents?
______	______	Do I have the knowledge and skills required for my job?
______	______	Do I continuously increase my knowledge and skills?
______	______	Do I know my strengths and use them effectively?
______	______	Do I know how to establish, maintain, and restore trust?
______	______	Do I consciously work to make trust happen?
______	______	Do I effectively communicate my track record to others?

54 Stephen Covey and Rebecca Merrill. 2008. The Speed of Trust. Free Press. New York, New York.

55 Jim Collins. 2001. Good to Great: Why Some Companies Make the Leap and Others Don't. Harper Business, New York, New York.

_______ _______ Do I focus my efforts on delivering results rather than being active?

_______ _______ Do I finish what I start?

_______ _______ Do I get results in ways that inspire trust?

Capability is critical for establishing credibility, which in turn inspires others to trust you. You can enhance your credibility by playing to your strengths and committing to a lifetime of learning that enables you to adjust to shifting circumstances that demand new skills.

Beliefs

Daniel Kahneman, 2002 Nobel laureate, said *For some of our most important beliefs, we have no evidence at all, except that people we love and trust hold these beliefs. Considering how little we know, the confidence we have in our beliefs is preposterous – and it is also essential.*

As characteristics reflect your values, values reflect your beliefs. Much of your belief framework was learned at an early age from your parents, and later in life from authority figures.

A belief is your opinion or conviction that can be held without contemplation or reflection. Confidence in the truth or existence of something not immediately susceptible to rigorous proof. For example, Copernicus believed that the sun was at the center of the universe, an idea at odds with the commonly held belief that the earth was the center of the universe.

Beliefs are a scheme for understanding, shaping, and predicting your environment.[56] They are shortcuts for quickly categorizing information while conserving energy for your demanding brain. Many beliefs are mundane, for example, the sun will come up tomorrow. Other beliefs have important

56 Ralph Lewis. 2018. What actually is a belief? And why is it so hard to change? Psychology Today. 7 October. https://www.psychologytoday.com/us/blog/finding-purpose/201810/what-actually-is-belief-and-why-is-it-so-hard-change.

consequences, for example denying climate change will lead to catastrophic impacts on the environment and human well-being and prosperity.

The efficiency of your beliefs comes with a tradeoff. You tend to jump to conclusions and prefer the familiar to the unfamiliar. You try to align new information with existing beliefs rather than alter or construct a new belief. You tend to reject evidence that contradicts your beliefs. And you try to rationalize your beliefs so you can maintain a consistent self-image.

Science requires a willingness to amend your beliefs. The scientific method is intended to systematically override beliefs and follow the evidence. Willingly revising existing beliefs, or constructing new beliefs, is how you progress in your pursuit of healthy lands and waters.

So, what are your beliefs concerning keeping lands and waters healthy? Reflection and contemplation are required. Think about a place you manage or enjoy. Alternatively, you could think about a broader challenge like pollution, food or water insecurity, biodiversity loss, or climate change. What beliefs do you hold about how to manage a place or resolve a challenge? Focus on the underlying beliefs or causes that guide your management, not the management actions.

As guidance, you can use an SAP approach, common among most contemporary philosophers who characterize beliefs as propositional (a proposition is an act of offering or suggesting something to be considered, accepted, adopted, or done). The *S* is the subject, *A* is the attitude (manner, disposition, feeling, or position), and *P* is the proposition. For example,

- I (*S*) believe (*A*) that human activities produce an overabundance of greenhouse gas emissions that accelerate climate change (*P*).
- I believe that habitat destruction and degradation lead to biodiversity loss.
- I believe that human well-being and prosperity require a resilient and sustainable planet.

- I believe that people will commit to healthy lands and waters in their own best interests and the interests of others.

Take your time with this exercise. Clarifying your beliefs informs your purpose.

Purpose

Now that you understand the characteristics you admire in others, and your values and beliefs, you are ready to complete the journey and discover your purpose. Your purpose is the reason you do something, the driving force for your determination and resoluteness, and the reason you get out of bed in the morning. Copernicus's purpose was to develop a model to support his belief that the sun was the center of the universe. A purpose he achieved despite significant persecution.

Purpose is your why. If you are clear about why you are doing something, all your decisions will be simpler. So too, if you can clearly state to others why you are doing something, others will understand the context of your decisions. A clear statement of why enables anyone in a working group, project team, or organization to make decisions.

How do you identify your purpose? Answering these questions will help. Begin by capturing what first comes to mind and then walk away. Return after doing something else and determine if your answers resonate with you.

- What do you most enjoy doing?
- What three things must you do every single day to feel fulfilled in your work?
- How would you spend your time if you didn't work?
- Thinking back on your life, what key themes help you understand what is important to you?

- When your life is ending, what will you regret not doing?
- What trends and patterns do you see around you?
- What message do you want people to take from you?
- What is next?

When you are comfortable that you understand what is most important to you, think about what the world would look like if you achieved your purpose. Write a vision statement, an outward expression of your purpose.

A vision encapsulates your optimal desired future state, a mental picture of what you want to achieve over time. Your vision should clearly and concisely reflect your values, beliefs, and purpose. Take your time, reflect, and edit until you create a statement that ignites your passion.

Here is my vision statement and examples from leading conservation organizations.

I envision a world where people understand the value of healthy lands and waters and make decisions that preserve those values and result in well-being and prosperity for all.

The Nature Conservancy envisions a world where the diversity of life thrives, and people act to conserve nature for its own sake and its ability to fulfill our needs and enrich our lives.

Conservation International imagines a healthy, prosperous world in which societies are forever committed to caring for and valuing nature, for the long-term benefit of people and all life on Earth.

The International Union for Conservation of Nature envisions a just world that values and conserves nature.

BirdLife International wishes to see a world where nature and people live in greater harmony, more equitably and sustainably.

You can deepen your understanding of purpose by developing a mission statement. Mission statements reiterate your why and clarify how. Another way to think about a mission statement is that it should contain some combination of challenge, action, service, and beneficiary. Try constructing a mission statement for yourself.

Here is my mission statement and some examples from leading conservation organizations.

I help responsible organizations achieve their missions through informative, persuasive, impactful, and evidence-based copywriting and leadership coaching.

The Nature Conservancy conserves the lands and waters on which all life depends.

Built upon a strong foundation of science, partnership, and field demonstration, Conservation International empowers societies to care for nature responsibly and sustainably, our global biodiversity, for the well-being of humanity.

The International Union for Conservation of Nature influences, encourages, and assists societies throughout the world to conserve the integrity and diversity of nature and to ensure that any use of natural resources is equitable and ecologically sustainable.

BirdLife International (i.e., The BirdLife Partnership) strives to conserve birds, their habitats and global biodiversity, working with people towards sustainability in the use of natural resources.

Your purpose, codified in your vision and mission statements, must remain front and center in your mind if you want to be satisfied with your life and lead others. Add them to your screen saver, print them out and place them over your desk, or carry them with you. Periodically review your vision and mission statements to determine if they still resonate with you and modify them as needed.

Beware of modifying your vision and mission statements to fit your current circumstances. Short-term alignment is misleading and will not bring you happiness, enhance your authenticity, or improve your leadership.

MAKE BETTER DECISIONS

In any moment of decision, the best thing you can do is the right thing, the next best thing is the wrong thing, and the worst thing you can do is nothing.

Theodore Roosevelt, 26th president of the United States

You make about 35,000 decisions a day.[57] If you exclude eight hours for sleep, that is more than 2,180 decisions per waking hour, 36 per minute. What to eat? What clothes to wear? What to study? What to prioritize at work? Who to assign to a project?

Each decision impacts not only your range of choices for subsequent decisions, but other people. As a leader, you must be aware of how your decisions affect the professional and personal lives of your followers.

The number of decisions you make is surreal. So how do you cope? For better or worse, you have default behaviors. You avoid making decisions. You are impulsive, choosing the first option presented to you. You defer to others, making the decision that best pleases those affected. You delegate decisions to friends and colleagues. And you weigh relevant factors, allocating your time and energy to the most consequential decisions.

Conscious decision-making and knowing your brain can help you make better decisions.

57 *Joel Hoomans. 2015. 35,000 decisions: The great choices of strategic leaders.* Leading Edge Journal. *https://go.roberts.edu/leadingedge/the-great-choices-of-strategic-leaders.*

Conscious Decision making

How do you minimize your bad decisions and maximize good decisions? Being conscious of your decision-making behavior is a good start.

Carl Jung determined that each person receives and processes information in four ways: thinking, feeling, knowing, and sensing.[58,59]

1. Thinkers use data, facts, and figures to analyze and infer. This is a conscious process.
2. Feelers consider their emotions. There is a sense of *feeling their way*. Feelings are the conscious consequence of emotions and emotions are unconscious.
3. Knowers are intuitive, perhaps not knowing how they made the decision, but they are sure it's right. The certainty that you are right does not mean that you are right. A sense of automatically *knowing the answer* is the conscious consequence of intuition, which is an unconscious process. Incidentally, intuition is a rational process, like thinking, but it is subconscious and therefore you are unaware of it.
4. Sensors use the input from their senses to inform them. This is a conscious process.

What do these processes look like in practice? Imagine, if you have been asked to save an endangered species. If you are a thinker, you are going to obtain every bit of data you can, analyze it, and arrive at a solution through reasoning. If you are a feeler, you may be dismayed by the impacts on the species and its habitat you observe, and call for a moratorium on land use.

If you are a knower, you may have already decided on a solution based on what you heard or read about the situation. And if you are a sensor, you may

58 Carl Jung. 1913. *Psychological Types. A contribution to the study of psychological types. Coll. Works, Vol. 6*. London: Routledge and Kegan Paul.

59 Work by Jung forms the basis of Myers-Briggs and many other personality metrics.

want to watch and wait before deciding, or if that is not possible, understand what others are doing in a similar situation.

You use all four processing modes but favor one. Which process do you favor? For important decisions, try to consciously be informed by all four processes. What are you thinking? What emotions are you experiencing? What intuitively seems like the right decision? What do your senses tell you?

Know Your Brain

Your brain is amazing, adept at controlling all bodily functions and interpreting information from the outside world. You have an enormous neural network, roughly 100 billion neurons capable of making 100 trillion distinct and elaborate connections with each other. Also, your neural network is plastic, capable of adopting new configurations. New ideas cause new assemblages of neurons to come together.

Your brain has four primary parts. Your brain stem is the earliest and simplest part of the brain, responsible for the heartbeat, blood pressure, and breathing. Your cerebellum is evolutionarily newer and controls balance, posture, and coordination. Both your brain stem and cerebellum are critical to your function, but neither plays a critical role in decision making.

Your limbic system encompasses several functionally and anatomically interconnected areas – the amygdala, hippocampus, hypothalamus, and thalamus – necessary for self-preservation. Conceptualized as the feeling and reacting brain, your limbic system regulates autonomic (working automatically without conscious effort) and endocrine (secreting internally to the blood or lymph) function in response to stimuli.

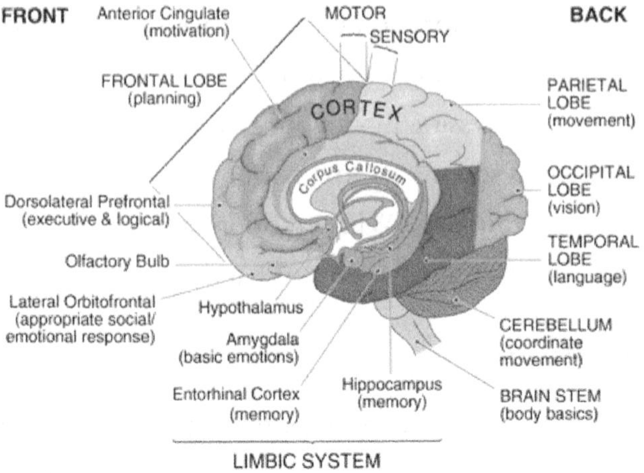

Your cerebral cortex is the newest and largest part of the brain and is responsible for most of your brain's function. The cerebral cortex has been described as the thinking brain.

Your cortex has four lobes – the parietal, occipital, temporal, and frontal – each divided into right and left hemispheres connected by the corpus callosum. Your parietal lobe is associated with movement, orientation, recognition, and perception of stimuli, your occipital with visual processing, and your temporal with perception and recognition of auditory stimuli, memory, and speech.

Your cortex's frontal lobe is associated with reasoning, planning, parts of speech, movement, emotions, and problem-solving. The prefrontal cortex is the only region of your brain able to take an abstract principle and apply it in an unfamiliar context to come up with an entirely new solution.

In general, the left hemisphere of your cerebral cortex controls speech, comprehension, arithmetic, and writing, whereas your right hemisphere controls creativity, spatial ability, artistic, and musical skills.

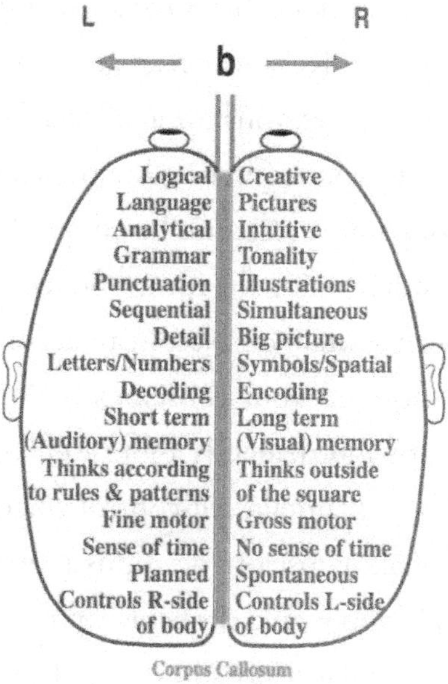

Brain Decision Making

Decision making is the domain of both your limbic system and the cerebral cortex. Emotions, managed by the limbic system, are a critical part of decision-making. Your every feeling is a summary of data, a visceral response to all the information you cannot directly access.

Your unconscious brain continually processes information, translating it into vivid visceral signals detected by your conscious brain, that is, the cortex. Your conscious brain can then add this information to the decision-making process, or if it chooses block information from the unconscious brain.

The advantage of emotional input is that your limbic system has been refined by evolution over the last several hundred million years. Dopamine helps regulate your emotions, and dopamine neurons generate patterns based on your experience. Your brain cells compare your expectations to your experiences, changing themselves if expectations are not met. Through endless testing, these flexible cells become the source of your expertise, allowing fast decisions based on little information. The best experts embrace intuitive thinking.

Expert or not, studies estimate that 80 to 100 percent of your decisions are driven by the limbic system. Reliance on an automated decision-making system is a blessing and a curse. Yes, you have a process for making decisions without having to check with your conscious brain each time. Your limitation is that the underlying dopamine system is based on pleasure now, overvaluing the near term while undervaluing the future. So too, your limbic system's reliance on experience can contain obsolete instincts in new circumstances.

Should you rely more on your conscious brain? Your cortex is already working at full capacity and demands a lot of energy to operate. Consider your ability to think and maintain self-control when your blood sugar level drops! If you tried to make conscious decisions about everything (remember, 35,000 decisions a day), you would probably lapse into a coma.

Save your cortex for what it does best – address abstract principles and novel situations. This feat is accomplished using your working memory, which keeps information in short-term storage so you can manipulate and analyze it. Unfortunately, you can only handle about seven pieces of information at the same time.

The reality is that your brain continues to evolve and adjust within your lifetime. Both your unconscious, emotional limbic system, and your conscious, rational cortex are essential to your decision-making. In fact, your decision-making is impossible without emotion.[60] Does your brain have limits? Yes, and familiarity with those limits helps you make better decisions.

Decision Making Limitations

What are those limits? For one, you are prone to overthinking. The part of your brain that monitors behavior, the prefrontal cortex, starts to interfere with decisions normally made without thinking. You become impulsive, begin to second-guess yourself, focus on information that does not matter, and rely on misleading shortcuts. Ultimately, you take longer to make worse decisions.

60 Antonio Damasio. 2005. Descartes' Error: Emotion, Reason, and the Human Brain. Penguin Books. New York, New York.

Trying to correct a bad golf swing or a batting stance are common sports examples of overthinking. But overthinking is not limited to sports. How much do you struggle trying to buy a car or television, decide what to wear to an event, or what to eat for supper? Worse, how difficult is it to develop a business strategy, assess the value of employees, or decide the priority for the day?

Your prefrontal cortex is also susceptible to meaningless anchors, reflecting our penchant for giving credence to irrelevant information. This is illustrated by the United Nations Game.[61] People were asked how many countries are in Africa after they witnessed the spin of a roulette wheel. People viewing higher roulette wheel numbers guessed a higher number of countries than people viewing lower numbers.

Another limitation to be aware of is the tension between different parts of your brain. Sometimes the tension is solely within the limbic system, reflecting conflicting emotions. At other times, the tension is between the limbic system and the cortex with the latter trying to resist impulses from the former. Your brain is uncomfortable with uncertainty and tries to end the debates within your limbic system or between your limbic system and cortex. Crucial information may be neglected, and bad decisions are the norm.

Make Better Decisions

How do you use consciousness and knowledge of your brain to make better decisions? You think about thinking. Here are some guidelines.

- Avoid making decisions when you are tired, stressed, or being influenced by people with an agenda.
- Try to name what you are deciding and the emotions the process is invoking.

61 Daniel Kahneman. 2011. Thinking Fast and Slow. Farrar, Straus, and Giroux. New York, New York.

- Be conscious of your decision-making process, i.e., thinking, feeling, knowing, and sensing. Try to blend the information from each behavior.
- Focus on your mistakes so you can consciously consider the errors internalized by your unconscious.
- Challenge your thinking. Ask what it would take to change your mind.
- Uncertainty is certain so embrace it. Difficult challenges rarely have simple solutions.
- Develop your ability to deal with competing viewpoints.
- Moral decision-making is about emotions and sympathy. Do not confuse moral decisions with rational thinking.
- Favor your conscious, rational cortex if facing
    - A novel situation.
    - A choice between immediate reward and long-term gain.
    - A relatively simple problem.
    - A decision of little consequence.
- Favor your unconscious, emotional limbic system if facing
    - A situation you experienced repeatedly in the past and to which you learned how to respond appropriately.
    - A complex problem.
    - An important decision.

The best leaders know how unsophisticated their automatic problem-solving skills can be when applied to complex problems. Be conscious of your decision-making process and consider the strengths and weaknesses of your brain. Don't hesitate to have your thinking and decisions be challenged by others.

EMBRACE CHANGE

Some people don't like change, but you need to embrace change if the alternative is disaster.

Elon Musk, technology entrepreneur

Change Happens

Leo Tolstoy, the Russian novelist, said, *Everyone thinks of changing the world, but no one thinks of changing himself.* If you want to lead and to promote change in others, you first must learn to manage changing yourself.

Change is constant, easily confirmed by reflecting on your life. Are you living the life you thought you would live? Is this the career you planned? Is this where you thought you would be living. Is this the partner you had imagined?

Can you avoid change? No. Leaders guide others through change. And to guide others through change you must embrace change yourself, to minimize its threats and maximize its opportunities.

Change is necessary. If your fisheries are depleted because the mangrove forest nursery was destroyed, then you need to eat something else or replant and sustain your mangroves.

If polar bears scavenge your trash for food because climate change-induced ice melt prevents them from hunting seals, then you need to protect yourself, move, or reverse climate change.

If you are lost, then you need to consult a map to find your destination. If the road you took is dissatisfying, you need to take another road.

You live in a rapidly changing world. Urbanization continues to expand as people migrate to cities for jobs and a higher quality of life.

Some agricultural practices are unsustainable, leading to the loss of soil fertility and casting doubt on our ability to feed a rapidly expanding population.

Renewable energy is a growing professional opportunity, an economic driver, and essential for avoiding a climate change catastrophe that will kill tens to hundreds of millions of people and diminish the health of our planet for many generations to come.

Adaptation is a critical life skill. The difference between embracing change and chasing it is the difference between a leader and a follower. Buddha said, *What you are is what you have been. What you will be is what you do now.*

Change is necessary when your expectations are at odds with your reality. Change happens with or without your consent. Charles Darwin, noted evolutionist, said, *It is not the strongest of the species that survives, nor the most intelligent, but the one most responsive to change.*

Growth requires you change the way you think, feel, behave, and act. Discovery, learning, and experimentation require change. Creating and innovating require change. Meeting new challenges requires things to be different than they are now. Making the world a better place requires change.

Resisting change is bad for your health. Emotionally you become easily agitated, frustrated, and sullen. You feel overwhelmed as if you are losing control. Relaxing is difficult, self-esteem diminishes, and you avoid others.

Physically, your energy diminishes. You get headaches, stomach upsets, and your muscles tense, causing aches and pain. You may experience chest pains, a rapid heartbeat, and insomnia. Colds and infections increase in frequency, and you may lose interest in sex.

Why is Change So Difficult?

If change is constant and necessary, and avoiding change makes you feel bad, why do you resist change?[62,63,64,65] Because change is different, at odds with your propensity for habits. Sometimes you prefer what you have rather than something different, even when what you have is dissatisfying. Even if you accept change as necessary you may still want to avoid the hard work of change.

Change may be a surprise triggering your sympathetic nervous system to respond with fight or flight. Uncertainty feels like you have lost control. When imposed on you by others, change can make you feel incompetent or obsolete. You fear that you may fail, lose status, your job, or a relationship. This fear may be exacerbated by a lack of trust in the competency or integrity of the change agent (e.g., your boss or partner).

Understanding Your Response to Change

You can learn to avoid viewing change as misfortune, lessening its pain and stress. You can learn to embrace change as an opportunity for growth, well-being, and prosperity.

Change tests your intellectual and emotional capacity. Understanding your values, purpose, and decision making are the bedrock for embracing change. Only you can know how to authentically move forward. You understand how your brain's decision making is biased toward resisting change, over-thinking, considering irrelevant information, and longing for certainty.

62 Rosabeth Moss Kanter. 2012. Ten reasons we resist change. Harvard Business Review. https://hbr.org/2012/09/ten-reasons-people-resist-change.

63 Ralph Ryback. 2017. Why we resist change. Psychology Today. https://www.psychologytoday.com/us/blog/the-truisms-wellness/201701/why-we-resist-change.

64 Robert Tanner. 2019. Organizational change: 8 reasons why we resist change. https://managementisajourney.com/organizational-change-8-reasons-why-people-resist-change.

65 Daniella Whyte. 2016. 5 reasons people resist change and what we can do about it. Inc.com. https://www.inc.com/daniella-whyte/5-reasons-people-resist-change-and-what-we-can-do-about-it.html.

The change curve model can help you develop the emotional capacity to embrace change.[66] Originally developed by Elizabeth Kubler-Ross to describe states experienced by terminally ill patients, the model was expanded to include any form of personal loss (i.e., change).[67] The change curve - encompassing some combination of shock, denial, anger, bargaining, depression, and acceptance - has been adopted by change management professionals to predict performance after the announcement and implementation of workplace change.

If you understand and recognize the change curve emotional states, you can gain some control over your response to change. So too, you begin to understand that what you are feeling is not unique but is a normal response to change. That said, not everyone will experience every emotional state. And the length of time in any state depends on individual capacity for and resistance to change.

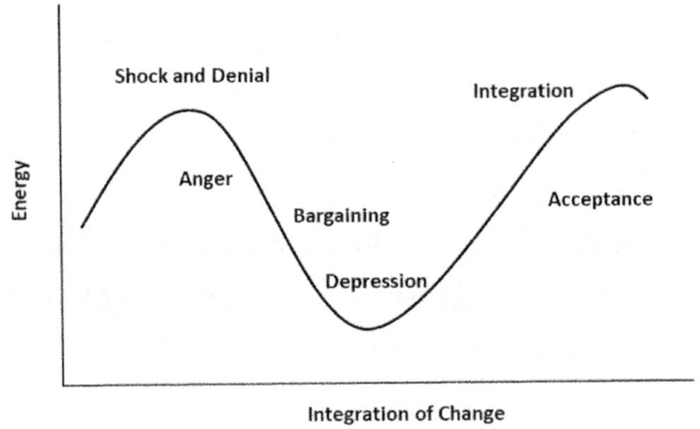

Source: Adapted from Elisabeth Kubler-Ross and David Kessler 2014

How do you recognize the emotional states triggered by change and their effect on your energy? Shock and denial are high energy states that trigger

66 Elisabeth Kubler-Ross. 1969. On Death and Dying: What the Dying Have to Teach Doctors, Nurses, Clergy, and Their Own Families. Scribner. New York, New York.

67 Elisabeth Kubler-Ross and David Kessler. 2014. On Grief and Grieving: Finding the Meaning of Grief Through the Five Stages of Loss. Scribner. New York, New York.

uncertainty, fear of incompetence, or no longer feeling needed. You cling to the status quo. Anger reflects a feeling of betrayal. You may try to bargain your way out of the change.

When you cannot bargain, depression arises, characterized by skepticism, frustration, and apathy. You isolate yourself from the source of the change and others.

Acceptance and integration reflect increasing energy and a focus on the future, accompanied by increasing optimism and enthusiasm. You recognize the opportunity presented to you, perhaps experiencing relief that tragedy did not occur. Hope and trust return.

Embrace Change

What else will help you manage change?[68,69,70] Accept that change is inevitable, uncertainty is common, and resistance is futile. None of these are inherently bad.

- Recognize the opportunity for renewal, and personal and professional growth.
- Take control of change by choosing to make change, rather than let change make you.
- Talk to others about your doubts and fears.
- Monitor your emotions to avoid self-loathing and fight or flight responses.
- Examine different perspectives, including that of the person initiating change, as a check against your emotions.

68 Nicolette Stinson. Undated. How to embrace any major life change. The Chopra Center. https://chopra.com/articles/how-to-embrace-any-major-life-change.

69 Katherine Russell Rich. 2014. 10 ways to embrace change. Real Simple.com. https://www.realsimple.com/work-life/life-strategies/ways-to-embrace-change.

70 Rhett Power. 2016. 5 ways to embrace change at work and in life. Inc.com. https://www.inc.com/rhett-power/5-ways-to-embrace-change-at-work-and-in-life.html.

Resilience is paramount. To flourish, to lead, you must develop the capacity to incorporate new ways of living and working. We all suffer setbacks, what is important is how well we recover.

PART 3
LEAD OTHERS

LEADERSHIP TYPES

Different times need different types of leadership.

Park Geun-hye, first woman president of South Korea

Leading others, even when you have authority, requires the development of skills and the application of practices. Simply telling people what to do seems simpler but it is not. People are far more productive, and the organizations they are part of far more successful, when employees are clear about their responsibilities and feel their contributions are valued. A supervisor unprepared to lead can significantly reduce employee engagement and organizational performance.

Everyone should be treated respectfully and held to the same standards. But leadership is not knowledge learning where you follow a set of established instructions down a single pathway to one conclusion. Leadership is not a recipe. Leadership is the authentic you. The *you* defined by your purpose, your approach to decision-making, and your ability to embrace change.

Each leader leads differently. What works for one leader may not work for another leader. So too, each follower has their strengths, weaknesses, motivators, and goals. Do not confuse authentic with unprepared. Understanding your behavioral tendencies, and those of others, is essential to being a successful leader.

Psychological Type

The theory of psychological type was introduced by Carl Jung, who categorized people into primary types of psychological function or preferences.[71] Jung proposed two perceiving functions - sensation and intuition - and two judging functions - thinking and feeling. The functions are modified by two attitude types - extraversion and introversion.

Jung's work forms the basis for several common personality typing instruments such as BEST (Boss, Engager, Supporter, and Technical), the Blake Moulton Management Grid, Myers-Briggs Type Indicator, and DISC (Dominance, Influence, Steadiness, and Conscientiousness).[72,73,74,75] Here we use the BEST Personality Indicator and the Blake Moulton Management Grid to illustrate the benefits of typing to leadership.

Boss, Engager, Supporter, Technical (BEST)

BEST, like many other typing models, is intended to help with hiring but is repurposed here to help you understand tendencies. BEST asks two questions. First, are you fast or slow when making decisions? Second, when working as part of a team, are you most concerned about the results or the people? By locating your answers along horizontal and vertical spectrums you can determine your tendencies, or through observation and review, those of others.

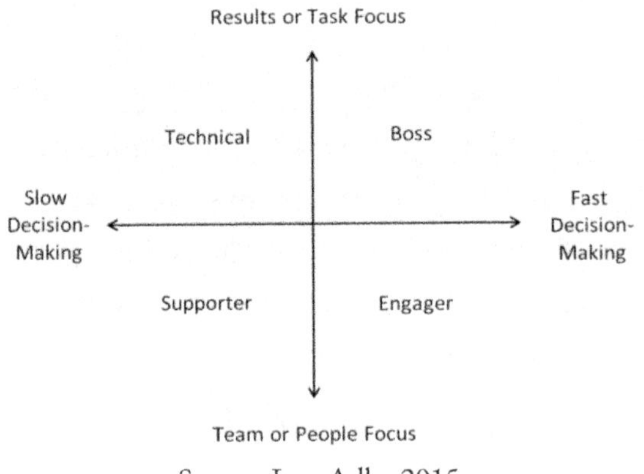

Source: Lou Adler 2015.

71 C.G. Jung. 1921. Psychological Types. Rascher Verlag. Zurich, Switzerland.

72 Lou Adler. 2015. Simply the BEST personality test. LinkedIn. 31 July. https://www.linkedin.com/pulse/simply-best-2-question-personality-test-lou-adler/.

73 Blake, Robert R., Jane S. Mouton, Louis B. Barnes, and Larry E. Greiner (1964), Breakthrough in Organization Development, Harvard Business Review, 42 (6), 133–155.

74 Isabel Briggs Myers. 2016. An Introduction to Type. 7th Edition. The Myers-Brigg Assessment Publisher. Sunnyvale, California.

75 Wiley. 2007. Everything DISC: Research Report for Adaptive Testing Assessment. John Wiley and Sons. Hoboken, New Jersey.

If you like to make quick decisions and focus on results, you may think of yourself as confident and driven. Others may perceive you as insensitive and bossy.

If you like to make quick decisions, but are team- or people-focused, you may think of yourself as persuasive and getting things done. Others may perceive you as disingenuous and rushed.

If you make slow decisions and are team or people focused, you may think of yourself as collaborative and thoughtful. Others may perceive you as sacrificing results for relationships.

And if you favor slow decisions and a focus on results you may think of yourself as analytical and deliberative. Others may perceive you as controlling and insensitive.

Blake Moulton Managerial Grid

The popular Blake Moulton Managerial Grid model, influenced by Fleishman, explicitly addresses results by replacing decision-making speed with concern for performance or production.[76,77]

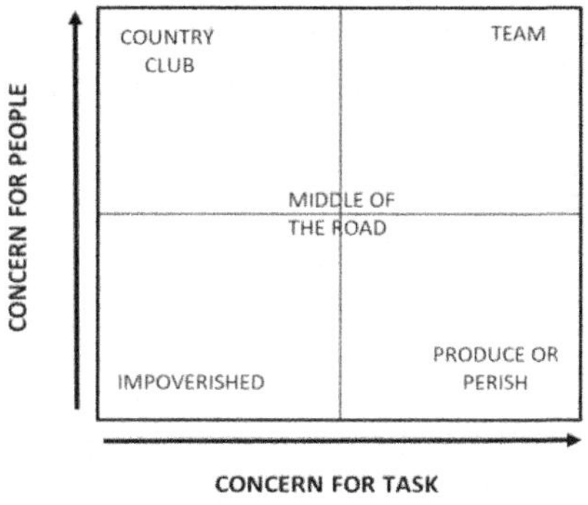

Source: Adapted from Fleishman 1957.

76 Fleishman E. A. (1957a), A Leader Behavior Description for Industry, in R. M. Stogdill and A. E. Coons (Eds.), Leader Behavior: Its Description and Measurement. Columbus, Ohio: Ohio State University, Bureau of Business Research.

77 Fleishman E. A. (1957b), The Leadership Opinion Questionnaire, in R. M. Stogdill and A. E. Coons (Eds.), Leader Behavior: Its Description and Measurement. Columbus, Ohio: Ohio State University, Bureau of Business Research.

The Blake Moulton Managerial Grid model identifies five kinds of leadership behavior.

1. *Country Club* leaders have high concern for people, and low concern for getting the work done. People feel good but fail to achieve.
2. *Impoverished* leaders have little regard for people or the work. These are not leaders at all, caring mainly about themselves and afraid of making mistakes.
3. *Middle of the Road* leaders try to balance concern for people with getting work done. An attractive idea in concept, these leaders tend to lack the authority and decisiveness to lead well.
4. *Produce or Perish* leaders are task-oriented and exhibit little concern for people. There is a tendency for dictatorship, which can be effective in a crisis but is not sustainable in the long term.
5. *Team* leaders blend high concern for people and a commitment to getting the work done through a collaborative teamwork approach. People feel respected, empowered, and are committed to achieving their goals. However, the approach is most effective when the followers are highly skilled in working in a team and suitably mature.

Blake and Moulton hold that *Team* leadership is the *one best way.* In many situations this may be true but reliance on a single type of leadership can limiting and maladaptive.

None of the typing methods are perfectly suited to assessing leadership types, potential, or effectiveness. As with any attempt to categorize, the methods provide a starting point for leadership type awareness. Any leadership type must recognize that they and their followers share a common purpose.

Lessons from Leadership Typing

Limitations aside, there are four leadership lessons to be learned from psychological typing. First, BEST, the Blake Moulton Managerial Grid, and other typing instruments reveal tendencies, not competencies. Your tendencies are not necessarily the same every day or in every situation. Do not take too much credit for your apparent strengths or feel too guilty about your apparent weaknesses.

Second, as you mature in your life and career, your behavioral tendencies become less extreme through trial, error, and training. You tend to migrate toward the middle. Consciously adapting the behavioral tendencies of others, particularly those of your diagonal opposite, is effective.

Third, BEST, the Blake Moulton Managerial Grid, and other typing instruments provide a construct for deciphering other's behavioral tendencies. This knowledge supports your efforts to coach an employee to improve their ability to work with you and others. Leaders do not use the knowledge to manipulate others.

Fourth, as a leader your goal is to become type adaptable. The best leaders read the room to understand the tendencies of others. Are there too many fast decision-makers or too few focused on the team? A leader recognizes these imbalances and assumes the under-represented role or roles.

BUILD TRUST

He who does not trust enough, will not be trusted.

Lao Tzu, ancient Chinese philosopher and writer

What is trust? Trust is a reliance on the character, ability, strength, and truth of someone. In a leadership context, you are asking someone to trust your integrity and competence. In the workplace, trust is analogous to driving someone home after a long evening out partying. They trust that you will get them home safely. At work they may not be risking their life, but they are trusting you with their livelihood. Trust is an awesome responsibility.

Is trust commonplace? If asked, *Do you think most people can be trusted?* what would you say? The results of a 2014-2015 survey by the World Values Survey Association and the European Values Study are mixed.[78]

Only 10 percent of Brazilians believe most people can be trusted. The Americans and the British are becoming less trustful, with only 30–35 percent responding positively in recent surveys. Other countries slipping include Russia, Mexico, and South Korea. If you relish trust, go to Scandinavia or the Netherlands, where 60 percent or more respondents think most people can be trusted.

A 2019 Pew Research Center study of trust in the United States found only 25 percent of adult Americans trust the federal government, and 64

78 David Halpern. 2015. Social trust is one of the most important measures that most people have never heard of – and it's moving. 12 November. https://www.bi.team/blogs/social-trust-is-one-of-the-most-important-measures-that-most-people-have-never-heard-of-and-its-moving/.

percent believe that low trust is an impediment to solving problems[79]. Only 22 percent of adult Americans have high trust for each other, while 35 percent have little trust for each other. Trust is lowest among minorities, younger, less educated, and the poorest Americans.

The Benefits of Trust

Is trust important? Unequivocally, yes. Level of trust predicts national economic growth, health, and life satisfaction.[80] Organizationally, trust is a strong predictor of employee satisfaction. Organizations with leaders who create a climate of trust outperform low trust organizations by 286 percent.[81] Trust is the number one differentiator between the top 20 and bottom 20 percent of companies evaluated for corporate innovation.[82]

Stephen Covey and Rebecca Merrill explain the impact, positive and negative, in simple terms.[83] If you have a low trust environment, everything takes longer. People examine the motives of others, trying to figure out the correct response or complaining to their colleagues. Since time is money, costs increase. In high trust environments, decisions are made and executed with minimal distraction and error in a timely manner. Costs are reduced.

Three Steps to Achieve Trust

You can build trust through a three-part process. First, trust yourself. Rely on your values, clarity of purpose, improved decision-making, and your willingness to embrace change (see chapters 3–5). Be authentic.

79 Pew Research Center. 2019. Trust and Distrust in America. July.

80 David Halpern. 2015. David Halpern. 2015. Social trust is one of the most important measures that most people have never heard of – and it's moving. 12 November. https://www.bi.team/blogs/social-trust-is-one-of-the-most-important-measures-that-most-people-have-never-heard-of-and-its-moving/.

81 Pamela Shockley-Zalabak, Sherwyn Morreale, and Michael Hackman. 2010. *Building the High-trust Organization: Strategies for Supporting Five Key Dimensions of Trust*. Jossey-Bass. San Francisco, California, US.

82 PricewaterhouseCoopers. 1999. Innovation Survey. London.

83 Stephen M.R. Covey with Rebecca R. Merrill. 2006. *The Speed of Trust*. Free Press. New York, New York.

Second, model the behaviors that inspire trust. Behaviors that reflect your values and the values necessary for a healthy and productive relationship.

Act with integrity

- Be sincere and honest in your words and actions.
- Keep your promises.
- Admit your mistakes and hold yourself accountable.
- Do what is right rather than be right.
- Act on good ideas no matter where they come from.
- Challenge outdated positions.
- Accept feedback from others.

Be Competent

- Do work that matches your skills.
- Continually improve and expand your skills.
- Deliver results.
- Know when to ask for help.

Be a Team Player

- Listen to others with patience and understanding.
- Trust others.
- Be transparent with knowledge.
- Show respect.
- Acknowledge help from others.
- Help others succeed.
- Share leadership.

If you model these behaviors, others will notice your success. As importantly, they will notice that you achieve success with integrity, fairness, teamwork, and humility. They will appreciate that you hold yourself accountable. So too, they will recognize your passion, your focus on the greater good, and your willingness to help others succeed. What you do is more important than what you say.

The third step in building trust is to set the expectation that the behaviors you model are not just your values, but the values that make the organization financially sustainable and mission impactful. Include creating trust in everyone's performance objectives. Get results by building a culture of trust, and holding everyone accountable for acting with integrity, competence, and as part of a team.

PROMOTE CHANGE

He who rejects change is the architect of decay. The only human institution which rejects progress is the cemetery.

Harold Wilson, former prime minister of the United Kingdom

Kouzes and Posner said, *leaders are men and women who guide others through adversity, uncertainty, hardship, disruption, transformation, transition, recovery, new beginnings, and other significant challenges.*[84] Followers like leaders who are willing to test bold ideas and take calculated risks. Proactivity consistently produces better results than reactivity or inactivity.[85]

Having developed authenticity, improved your decision-making, embraced change, and built trust you can tackle organizational change. You learned that change is constant, necessary, inevitable, and a test of your intellectual and emotional capacity. You learned that understanding change will help you embrace change. Now, as a leader, you need to help others understand and embrace change.

The Unifying Principle of Transformation

To help you promote change, let's add one more reason why change is necessary - organizational life or death. Dr. George Land was a general systems scientist who studied creative performance. One

84 Jim Kouzes and Barry Posner. 2012. The Leadership Challenge. Jossey-Bass (Wiley). San Francisco, California. 25th Anniversary Edition.

85 J. Michael Crant. 1995. The Proactive Personality Scale and Objective Job Performance Among Real Estate Agents. Journal of Applied Psychology 80(4): 532–537.

seminal contribution was integrating creativity, growth, and change in the Unifying Principle of Transformation.[86] A principle applicable to not only to organizations, but to cells, organisms, and individuals.

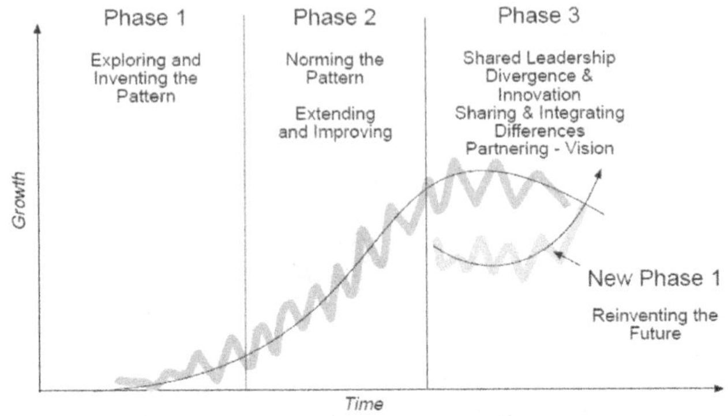

Source: George Land. 1973.

The unifying principle of transformation depicts organizational growth as an S curve with three distinct phases. In Phase One, the organization is a start-up or responding to external forces. It is in a state of chaos, exploring to determine what is possible. As the organization overcomes challenges and clarifies opportunities, it begins to form a pattern.

The pattern allows the organization to move into Phase 2 – growth through limitations. It eliminates unproductive behaviors, concentrates on what works, and begins to grow incrementally. Organizational life is seemingly controllable and comfortable.

At some point new challenges like competition or a shift in user needs force Phase 3 – radical change. The organization must return to Phase 1, reinventing itself to remain relevant. The alternative is increasing obsolescence and failure.

Phase 3 and a return to Phase 1 is a critical time in the health of an organization. The organization must learn fast from its mistakes. Practices,

86 George T. L. Land. 1973. Grow or Die: The Unifying Principle of Transformation. Dell Publishing Company. New York, New York.

products, and services need to be reassessed. Those no longer needed must be eliminated. New practices, products, and services with higher customer value and return need to be developed and offered. Change is paramount.

Phase 3 is also a critical time in the health of an individual. You must learn fast from your mistakes, re-assess your priorities and decisions, and eliminate those that no longer benefit you. You need to develop new priorities and behaviors. Again, change is paramount to avoid obsolescence.

Resistance to Change

Change is unavoidable so why do many people resist? Our brains hate change, preferring reward and certainty, avoiding threat and pain, and minimizing energy use. Change is also at odds with our propensity for habits. Change can trigger feelings of loss or obsolescence.

Those you want to lead have cognitive biases that make change difficult. You, and they, will overestimate the extent that you share beliefs.[87] You will assume that others possess the knowledge to understand what you are communicating.[88]

Like many leaders, you will treat change as a technical challenge when evidence and facts are secondary to emotions.[89] You and those you lead use mental models to help make sense of complexity.[90] Mental models reflect our views of the world, ourselves, our capabilities, and the tasks that we are asked to perform.

Our mental models are biased by brain receptivity to information that confirms preconceived notions, and our disdain for information that contradicts those notions. When you and those you lead disagree, your brains immediately became less sensitive to the information presented by the other.[91]

87 Tessa Basford and Bill Schaninger. 2016. The four building blocks of change. McKinsey Quarterly. April. https://www.mckinsey.com/business-functions/organization/our-insights/the-four-building-blocks--of-change.

88 Jane Kennedy. 1995. Debiasing the Curse of Knowledge in Audit Judgment. The Accounting Review 70 (2): 249–273.

89 Tali Sharot. 2017. The Influential Mind: What the Brain Reveals About Our Power to Change Others. Henry Holt and Company. New York, New York.

90 Dedre Gentner and Albert Stevens (Eds.). 1983. Mental Models. Lawrence Erlbaum Associates. Mahwah, New Jersey.

91 Tali Sharot. 2017. The Influential Mind: What the Brain Reveals About Our Power to Change Others. Henry Holt and Company. New York, New York.

Change also has a cultural component, reflecting an individual's relationship with others and the organization, and individual and collective assumptions, values, and beliefs. Individuals challenged to change their mental models consider not only their beliefs, but their role and relationships in the workplace.

Before tackling how to overcome resistance and bring about change, let's review the emotional states of the Kubler-Ross change curve.[92] This time view it with empathy, adopting the perspectives of those you want to lead.

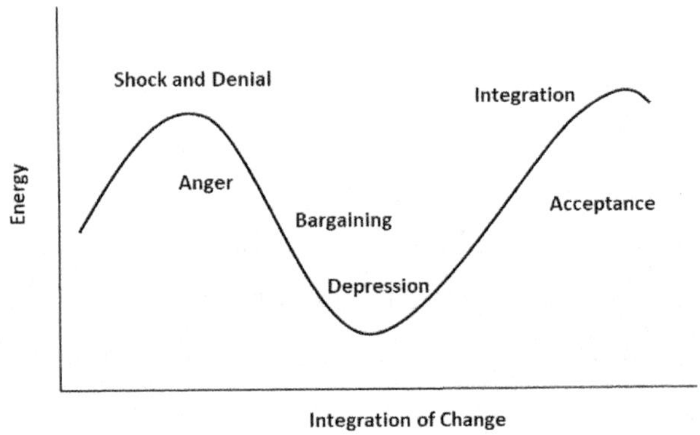

Source: Adapted from Elisabeth Kubler-Ross and David Kessler. 2014.

Your leader has signaled change is coming. How do you respond? Do you question why change is needed? Are you angry that the job you worked so hard at may change and you will have to learn a new job? Or perhaps you think you'll be fired.

Do you promise to do a better job to avoid change? Are you willing to accept that the organization must change, but want your role to remain the same? Do you accept that change will affect you, assume it will be bad, and become sullen and unproductive? What would convince you to accept change?

92 Elisabeth Kubler-Ross and David Kessler. 2014. On Grief and Grieving: Finding the Meaning of Grief Through the Five Stages of Loss. Scribner. New York, New York.

Shock, anger, bargaining, and depression are common responses to change. Many people say they like change, but often what they mean is they like change they can control.[93] When someone else initiates change, they imagine change for the worse not change for the better. Rare is the individual willing to sacrifice themselves for the betterment of the organization.

Bringing About Change

Successful change is not a marketing effort or a pep rally. It is not a reaction to a crisis, or fear of lagging behind your competitors. Successful change is not incentivized by compensation, or achieved through acquiring other assets. It is not new technology or any other quick fix. Change is a long, steadfast process, fraught with uncertainty, and made successful by a skilled leader and committed followers.

Flywheel, Hedgehog, and Discipline

Jim Collins of *Good to Great* fame attributes change at great companies to a combination of the flywheel effect, the hedgehog concept, and discipline.[94]

The flywheel effect is a gradual, consistent commitment to excellence. The flywheel – your organization – is at a standstill. It takes tremendous effort to get it moving in a new direction. Continued effort makes it move faster and faster. Someday, no one knows when, momentum begins to work in your favor and the flywheel accelerates.

The second change attribute is called the hedgehog concept. According to the Greek poet Archilochus, the fox knows many things, but the hedgehog knows one big thing. The hedgehog is a hardworking, no-nonsense, disciplined leader who can reduce complexity to a single, organizing idea that guides all decisions.

Hedgehog concepts are the result of asking and answering hard questions. Why are we failing? What can we excel at? How can we be financially

93 Aad Boot. 2017. Stop believing these persistent myths about leading change. LinkedIn.com/pulse/stop-believing-persistent-myths-about-leading-change-aad-boot. 5 June.

94 Jim Collins. 2001. Good to Great: Why Some Companies Make the Leap and Others Don't. Harper Business, New York, New York.

sustainable? What are we passionate about? What do we need to stop doing? Collins's great companies required an average of four years to solidify their hedgehog concepts. Change takes time.

Discipline is the final attribute of organizations that go from good to great. Great organizations did not start with where they are going, but with who. Think about it for a moment. When change is required, who among us is prescient enough to know what change will look like? As Collins reveals, great companies start by identifying the smart, open-minded, agile, innovative, hardworking people needed to complete a lengthy and grueling exploration leading to change.

The Influence Model

McKinsey & Company's perspective is that change is most successful when organizations use an influence model with four key actions to change mind-sets and behaviors.[95,96,97]

Fostering Understanding and Conviction focuses on why change is happening. This surfaces any disconnects between the change leader's and the followers' beliefs and knowledge.

Reinforcing with Formal Mechanisms aligns organizational structure and processes with change. Expectancy theory predicts that individual behavior is motivated by the resulting performance, and performance will be rewarded.[98]

Developing Talent and Skills affords current staff the opportunity to learn new skills needed for the changed workplace. An organization's willingness to develop existing staff alleviates fears while reducing the expense of firing and hiring.

95 Emily Lawson and Colin Price. 2003. The psychology of change management. The McKinsey Quarterly. https://www.mckinsey.com/business-functions/organization/our-insights/the-psychology-of-change-management.

96 Tessa Basford and Bill Schaninger. 2015. The science of organizational transformations. The McKinsey Quarterly. September.https://www.mckinsey.com/business-functions/organization/our-insights/the-science-of-organizational-transformations.

97 Tessa Basford and Bill Schaninger. 2016. The four building blocks of change. April. The McKinsey Quarterly. https://www.mckinsey.com/business-functions/organization/our-insights/the-four-building-blocks--of-change.

98 Victor Vroom. 1995. Work and Motivation. Jossey-Bass Publishers. San Francisco.

Role Modeling behaviors and practices signals expectations to followers. Change leaders must model appropriate behaviors and should encourage other leaders in the organization to do so.

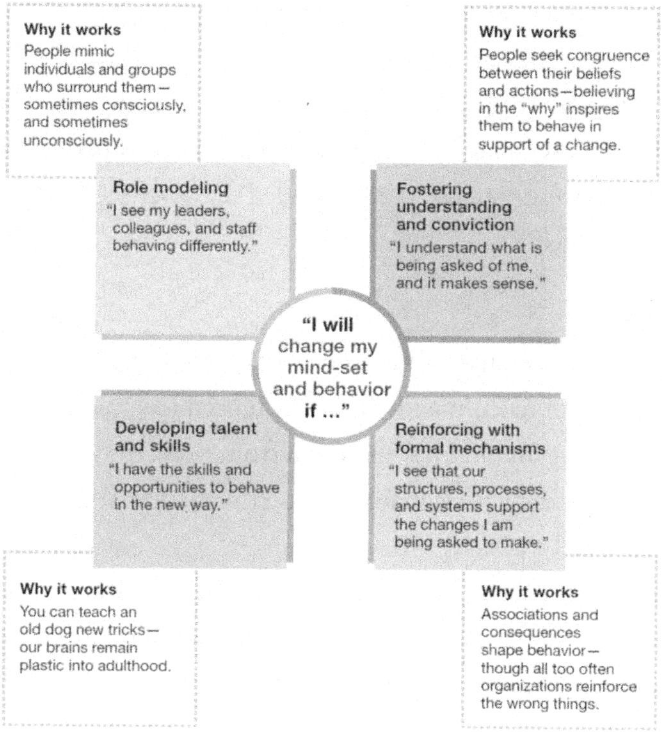

Source: Tessa Basford and Bill Schaninger. 2016.

Successful change requires acceptance at all levels of the organization.[99] You must clearly and consistently communicate that change is coming and how the process is progressing. Those you lead must understand their roles and responsibilities. And you and other change leaders must signal their commitment to change by being deeply and visibly involved in the process.

99 Dana Maor, Angelika Reich, and Lara Yocarini. 2017. McKinsey & Company. https://www.mckinsey.com/business-functions/organization/our-insights/the-people-power-of-transformations.

Change Personalities

A challenge to change is gaining acceptance across a group of individuals who view change differently. You must look for commonalities and employ empathy to identify individual mental models, to discover what people already believe and what they want to believe. And you must acknowledge the universality of the need to feel in control, to have hope for the future, and a longing to be part of a group.[100]

Kevin Eikenberry identified eight change personalities to facilitate managing disparate individuals.[101] I categorized Eikenberry's personalities as advocates, on the fence, or resistors and amended the descriptors and management recommendations.

Advocates

- *Eager Beavers* have little to no resistance to change. They can help you build momentum with others. You need to equip eager beavers to be change agents and restrain them from proceeding too quickly.
- *Evangelists* have a valuable passion for change. You need to encourage them to be patient with others.
- *High-Pressure Salespersons* believe in the change. You need to counsel them to engage others to show them why they are excited about the change.

On the Fence

- *Patient Pacifists* will not be early adopters, but they will not resist. You need to give them information and examples of others who have succeeded. They will become more willing to accept change as others come on board.

100Tali Sharot. 2017. Want to change minds? Try this. September 14. https://www.cnn.com/2017/09/14/opinions/the-influential-mind-sharot-opinion/index.html.

101 Kevin Eikenberry. No date. Leadership and Learning. https://blog.kevineikenberry.com/coaching-developing-others/successfully-leading-change-personalities-team/.

- *Fact Lovers* place the most value on the data and logic behind the change. You need to provide them detailed, relevant information, answer their questions, and let them share their findings with others.
- *History Professors* will know the past relating to this change. You need to let them share their experience and then focus them on how the lessons from the past will inform the future.

Resistors

- *Alarmists* see the problems with the coming change. You need to allow them to assess risk, and not dismiss their concerns. Address the concerns they raise before they affect others. Alarmists can become strong advocates if their concerns are addressed.
- *Ostriches* ignore or deny that the change is coming. You need to help them see the change as real, ask for their concerns, and then help them envision how the organization will be better once the change has been implemented.

The change personalities inform your understanding of people's perspectives, including their hopes and fears. As a leader, you must respectfully guide all change personalities through a collaborative approach to a productive consensus.

As the leader promoting change, the impetus is on you to create a healthy and safe environment.[102,103] Lead with enthusiasm. Highlight the consequences of not changing. Normalize resistance to change.

Theory of Change

The Kubler-Ross Change Curve taught us that many people struggle with change. They experience shock, denial, bargaining, and depression before accepting and becoming part of change.

102 Jim Kouzes and Barry Posner. 2012. The Leadership Challenge. Jossey-Bass (Wiley). San Francisco, California. 25th Anniversary Edition.

103 Aad Boot. 2017. Stop believing these persistent myths about leading change. June 5. https://www.linkedin.com/pulse/stop-believing-these-persistent-myths-about-leading-change-aad-boot/.

The theory of change process can help you lead people through the emotional states associated with change. Theory of change is a description and illustration of how your organization will achieve its desired impact and outcomes. The benefits of a Theory of Change exercise include:

- Enhanced clarity about desired impact and outcomes.
- Construction of a path to achieve the desired impact and outcomes.
- Articulation of assumptions that can be tested and measured.
- A shift in mindset from what the organization is doing now to what it needs to do.
- Agreement among stakeholders about what success looks like and how to get there.
- A powerful tool for communicating with others.

Theory of change is scalable and applicable across organization, program, and project levels.

Pragmatically, theory of change is a transparent, inclusive, and participatory exercise that identifies your organization's long term goal, and then works backwards to identify the preconditions that must occur to achieve the goal. What approaches will you take? How does each approach contribute to your desired impact or outcome? What are the essential elements of each approach? How do your approaches relate to each other? How will you re-allocate resources to support each approach?

The product of a theory of change exercise is illustrated in an outcome framework. My typical theory of change exercise has 6 steps.

1. Identify your organization's desired impact or outcome. What would the world look like if you achieve your desired impact or outcome? Who will be impacted? How will they be impacted?
2. Work backwards to preconditions necessary to achieve your impact or outcome, and preconditions for your preconditions, until you arrive at your current position.

3. Articulate the assumptions attributed to each precondition to ensure agreement for planning and action.
4. Identify relationships, if any, among pre-conditions in different paths.
5. Establish the means for measuring and assessing performance toward impact and preconditions achievement. Indicators are the variable being measured, population is the group you are measuring, and threshold is the minimum for the precondition to be successfully achieved.
6. Articulate a brief narrative explaining your theory of change.

A theory of change exercise is an iterative process. Continue iterating until consensus is achieved on a logical framework and assumptions, pre-condition relationships, and measures of success are documented.

Let's look at an example of how to use theory of change. In this example, a landowner wants to sustain their land and water.

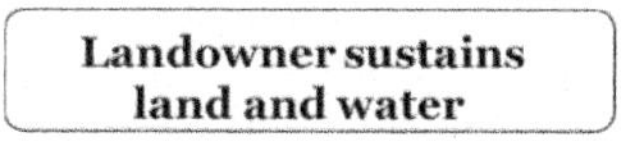

What is a precondition to sustaining land and water? The landowner must implement a management plan.

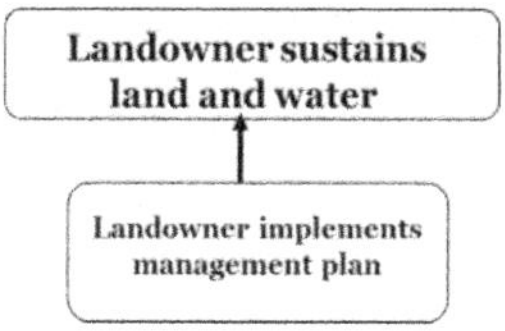

A precondition for implementing a management plan is to develop a management plan. For example, the landowner could ask a University Extension Agent to prepare the management plan.

What is a precondition to the extension agent's development of a management plan? The agent needs a vision and strategy to guide the management plan.

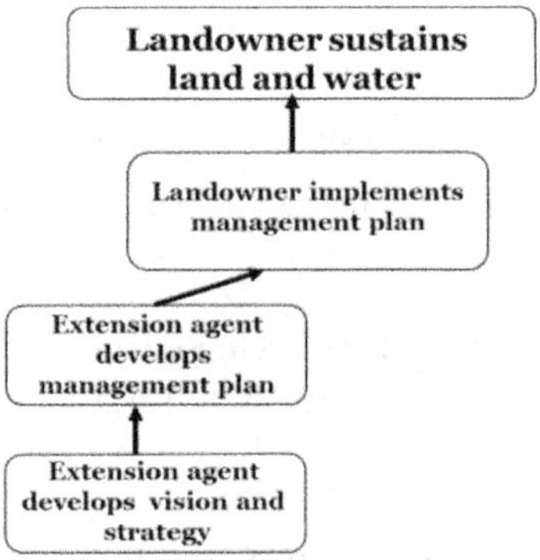

What is a precondition for developing a vision and strategy for the land and water? The extension agent will need to assess the value of the land and water. Also, the agent is going to need to talk with the landowner about what is important to them.

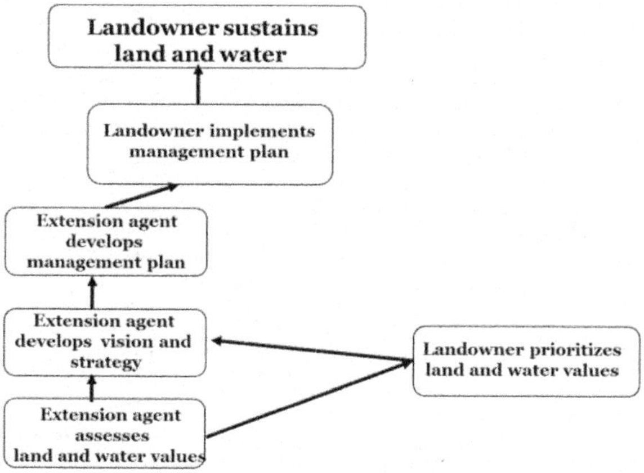

What is the precondition to assessing the value of the land and water? The extension agent must be hired by the landowner.

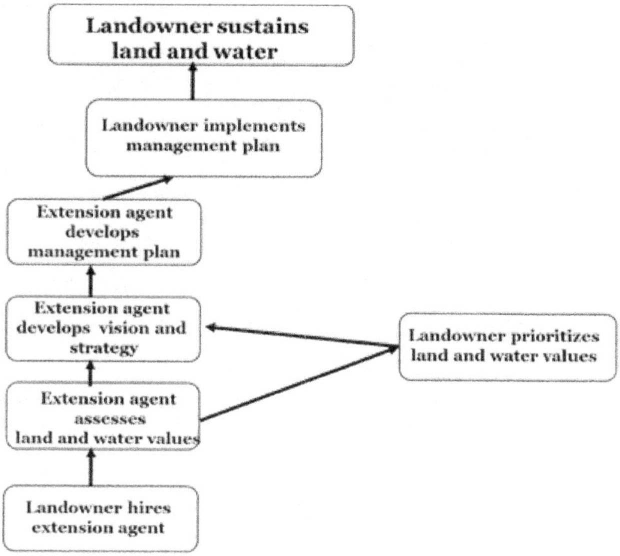

An alternative to hiring an extension agent is for the landowner to do the work themselves.

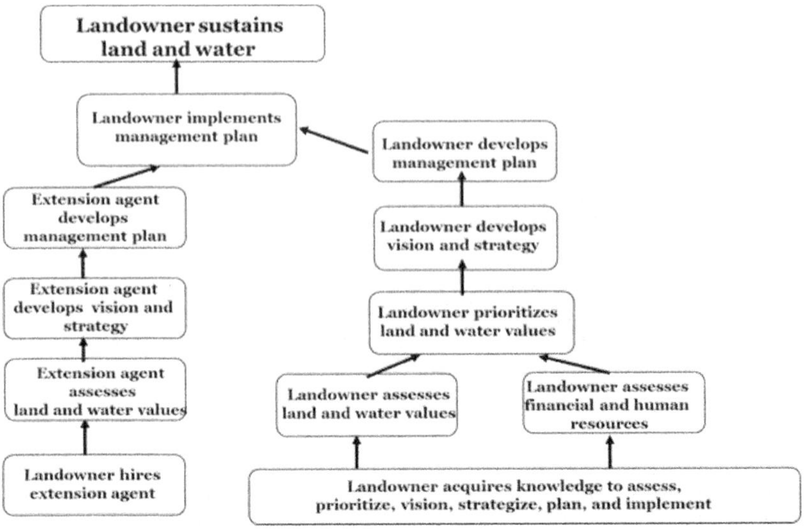

The exercise is not complete until assumptions, precondition relationships, and measures of success are articulated. For example, the alternative theory of change assumes that the landowner can acquire the knowledge to assess, prioritize, strategize, plan, and manage their land and water. Access to information, previous experience, and time are factors.

As to precondition relationships, even if the landowner chooses to hire an extension agent, the agent must consult with the landowner and the landowner needs to make a significant decision.

The ultimate measure of success is the sustainability of the land and water. Depending on the vision, strategy, and plan for the property, metrics could include adequate water to nourish crops, clean water to support fisheries, or biodiversity characteristics of the habitat.

One clarifying narrative is that the landowner will hire an extension agent to assess the values of their lands and waters. In consultation with the landowner, the assessment will inform the development of a vision for the land and a strategy for achieving the vision. The vision and strategy will inform a management plan to be implemented by the landowner.

Guidance for Promoting Change

Promoting change is hard. Change requires overcoming stakeholder cognitive biases, mental models, and cultural relationships. Change requires building stakeholder emotional capacity. Understanding change personalities and implementing a respectful, collaborative, inclusive change process will help.

As the leader promoting change, the impetus is on you to create a healthy and safe environment. Lead with enthusiasm, highlight the consequences of not changing, and normalize resistance to change.

Be clear about the intended impact of the change and define desired outcomes. Accept that the process may be more improvisational than scripted. Guide the conversation, allowing stakeholders to raise questions and find answers. Break big problems into small achievable actions. Generate and celebrate small wins.

Build psychological hardiness - the ability to cope with stress - by helping staff commit to the change process, take control, and see change as opportunity. And above all else be authentic, transparent, and trustworthy.

CREATE A SHARED VISION

In order to serve its purpose, a vision has to be a shared vision.

Warren Bennis, pioneer of the contemporary field of leadership studies

People who look forward, envisioning what could be instead of what is, are visionary. Vision is a key attribute of leaders. The Leadership Practices Inventory, a 360-degree assessment of leadership effectiveness based on more than 3 million respondents, found that after honesty, vision is the most desired attribute of a leader. Ironically, only 3 percent of a typical business leader's time is spent envisioning the future.

As a leader, Your responsibility as a leader is to think as far into the future as possible, amending your thinking as the future unfolds. Leaders with vision attract followers, inspire more effort, promote teamwork, mobilize collective action, and achieve better individual and organizational outcomes.

Earlier you learned how to create vision and mission statements to support your life's purpose. The exercise was deeply personal, requiring significant thoughtfulness and self-reflection. Do you ask your employees, board members, partners, and supporters to accept your vision and mission statements? You could. You might even succeed if they do routine work and only care about compensation. But that is not leadership and will not produce the level of stakeholder commitment needed to achieve the vision.

Alternatively, your vision could be so awe-inspiring and appropriate that everyone embraces it and follows your directions. To coin a phrase, the odds of this outcome are zero to none. A key to leadership is accepting that you are not prescient, and that you and your organization benefit from diverse ideas.

The more likely outcome of forcing your vision on others is that your best people resign, the remainder are demotivated, and you spend your days telling people what to do.

The Benefits of a Shared Vision

If you want a creative, motivated workforce, and don't we all, then a shared vision is essential. Imagine yourself trying to lead an organization in a visioning of the future. An organization where many of the people have their own purpose, and their own vision of the organization. Those staff, board, partners, and supporters want a vision that reflects their aspirations. Their vision and yours must be aligned in a singular organizational vision.

Your primary goal for a shared vision is to establish a direction for the organization that is compelling and sustainable. Compelling in the sense that staff, board members, partners, and supporters are inspired and supportive of what the organization wants to achieve. And sustainable in the sense that the vision has a substantial, long-term view and can reasonably attract enough income to offset expenses. A good vision can be a catalyst for change, growth, and success, and a yardstick for measuring progress.

An effective visioning exercise clarifies purpose for your staff, board members, partners, and supporters. The actions of each, especially staff, are enabled by using vision, mission, and strategy to simplify complexity, guide decision-making, set expectations, and prioritize activities. Staff can be connected to meaningful outcomes so that they understand the import of their, and their colleagues', activities, creating a culture of shared ownership and achievement. A common vision and connectedness promote workplace unity, raise morale, and facilitate collaboration.

Your vision also has external benefits. Marketing and communication will have narrative continuity. Your audiences will know who you are and what you are trying to achieve. This includes partners and donors. Also, a good vision helps you attract and retain employees, volunteers, and partners.

Preparing to Create a Shared Vision

Trust is an essential ingredient for creating a shared vision. A visioning process requires frank, respectful discussions with contributions from all stakeholders. If your staff, board members, partners, and supporters are not ready to trust, review the chapter on building trust. Schedule meetings with key individuals and talk about your commitment to transparency, honesty in words and actions, and the value you place on their contributions to the organization.

Leading a visioning process is hard work and a significant investment in time and resources. Many loathe prioritizing a shared vision. There are so many tangible challenges to tackle. Challenges with seemingly more immediate return on investment. Accomplished managers are not always adept at, and comfortable with, the active listening, sincerity, nonjudgment, patience, tolerance, and compassion needed to create a shared vision.

Managing the shared vision process requires you to embrace a mix of type A and type B personality traits. From type A, outgoing, ambitious, organized, proactive, and concerned with time management. From type B, working steadily, tolerant, circumspect, contextual, encouraging teamwork, and exercising patience in decision-making. Rest assured, blending the two personality types is possible, and essential for leadership. With a side benefit of reducing your risk of heart disease.

A shared visioning exercise can leave some disappointed that the organization's vision is not their vision. You can make stakeholder disappointment more bearable by accepting that the goal of a shared visioning process is not to please everyone.

If you could achieve unanimity, you have a problem with intellectual diversity, an attribute that diminishes your organization's creativity and innovation. Without creativity and innovation, you cannot react to inevitable change, and organizational failure is the ultimate outcome. Accept a little disappointment and help those who feel left out to see their critical role in the organization.

How to Create a Shared Vision

The process of creating a vision is as important as the vision. The shared experience is a chance to be heard and valued, a chance to take ownership, and a chance to change the way things are. All this helps build the team you will need to execute the vision.

What does a good vision look like? A good vision speaks not about your organization but about those you serve. It paints a clear and compelling picture of the future, one that is memorable and easily communicated to others. The vision is sufficiently specific that everyone understands what success looks like. And lastly, people will accept the vision as achievable. No one likes to chase an ideal they cannot attain.

Start by preparing yourself. Your overarching goal is to produce a vision that drives the mission, strategy, and actions of the organization. Theoretically, you are starting with a blank slate. Realistically, that slate already has some notes. They might reflect partner skills and experience if you are starting an organization. Or they reflect the current or easily attainable resources of your existing organization. Be cognizant of your, and participants', biases and try to keep an open mind.

The goal of the shared process is to generate ideas that will inform the vision and help you understand stakeholders' perspectives. Be prepared to listen deeply and sensitively, connecting to what is meaningful to others. Accept that people will disagree, but recognize that perspectives misaligned with the final vision will need to be addressed to maintain the team. Inspire confidence and creativity. Express positive emotions about the vision and the future. Commit to helping people be comfortable with change.

Decide who should be involved. Anyone responsible for the success and health of the organization should be included. At a minimum, this includes staff, board members, partners, and supporters.

Whether you meet with these groups together or separately depends on size, intergroup dynamics, and your skill at facilitating a productive discussion. I recommend separate conversations with people who are not stakeholders per se, but respected for their objectivity, wisdom, and authenticity. They often are less biased, ask the most provocative questions, and provide the most innovative insights.

Facilitator or no facilitator, that is the question. A skilled facilitator can plan group meetings, objectively ensure balanced participation and idea generation, and synthesize and share outcomes. Some stakeholders are more comfortable talking to an outside facilitator.

What an outside facilitator cannot do is offer timely guidance to the conversation, sustain trust, or assume the decision-making role. If you have facilitation skills, you are in the enviable position of going it alone or hiring a facilitator. In my experience, stakeholders are more interested in whether they had an opportunity to contribute, and if their ideas are reflected in the final vision, than in who facilitated the process.

To begin your conversations with stakeholder groups, share the process, expected dynamics, encouragement, and guidance. Here are some recommendations.

Process

- Explain who you will be speaking with, when, and for what purpose.
- Clarify that the group is gathered to help shape the vision, not create the vision.
- Explain that the goal of the group is to generate ideas.
- Name who will make the final decision. Typically, this is the Chief Executive Officer in collaboration with the Board of Directors.

Dynamics

- Insist that one person speaks at a time.
- Ask everyone to listen actively to others and repeat back to ensure understanding.
- Ask everyone to listen with the sensitivity to understand what is meaningful to all participants.
- Keep the environment positive. There should be no personal criticism.

- Participants should speak when they have something new to add to the pool of knowledge.

Encouragement

- Assert that change is OK. Creativity is encouraged. What we did before or are doing now may not be the best thing we can do tomorrow.
- Everyone's perspective is important regardless of their role in the organization.
- Collective intelligence is greatest when everyone contributes in equal measure. Do not dominate the discussion or remain silent.

Guidance

- Remind the group to stay user focused. This is about who we serve and how we serve them.
- Ask them to try presenting their ideas as stories. Who has a need, what is the need, and how can it be addressed?
- Ask what the future looks like if this was a picture.

Engaging Stakeholders

You may find it helpful to ease your stakeholders into the conversation. For example, you could ask them what the organization's core values are, why the organization exists, and what success looks like. Then pose three seemingly simple questions.

1. Who are we serving?
2. What problem are we trying to solve?
3. What would the world look like if we solved this problem?

To maximize learning, divide stakeholders into groups and meet with each group individually. For each group of stakeholders, address the three seemingly simple questions one at a time ensuring that every stakeholder contributes in relatively equal measure. Take accurate and unbiased notes with comments attributed to individuals. Thank them for their time and contributions and remind them of the process. Provide participants with unattributed notes from all stakeholder meetings.

Draft a vision statement when you finish talking with all the stakeholders. The statement should be concise, easy to understand, in the future tense, and describe the future the organization will work toward. Be clear about the benefits the organization offers and to whom. For example, (Organization) envisions a world where people understand the value of healthy lands and waters and make sustainable decisions leading to well-being and prosperity for all.

Lastly, does the vision statement excite you? Will it excite others? Your organization's vision needs to inspire staff to work and your board members and supporters to invest in the organization.

Regroup with your stakeholders and ask for their thoughts on the draft statement. Reactions will likely be mixed. Use this second meeting to give people a chance to make their case for revisions. Avoid wordsmithing and focus on the three essential questions:

1. Who are we serving?
2. What problem are we trying to solve?
3. What will the world look like if we solve the problem?

Remind them that the goal is to achieve reasonable consensus on the purpose of the organization.

Craft a final vision statement when everyone has had an opportunity to add their ideas to the discussion and you believe you have reasonable consensus. Meet with your stakeholders to unveil the vision statement and explain your thinking. Unless someone has an especially helpful insight, avoid modification. Perfection is the enemy of progress. The process was important. The vision can be modified as the work unfolds.

Add a Mission Statement

Creating a vision for the organization, the organization's *why*, is a critical first step. While you have momentum, continue the engagement process and construct a mission statement. The mission statement describes how the organization will achieve its vision.

At a minimum, a mission statement describes an action, a service, and a beneficiary. For example, (Organization) promotes scientific literacy and environmental stewardship through leadership, partnership, research, and education to ensure sustainability, well-being, and prosperity for all. *Promotes scientific literacy and environmental stewardship* is the action; *through leadership, partnership, research, and education to ensure sustainability, well-being, and prosperity* is the service; and *all is the beneficiary*. As with the vision, be clear and concise.

MOTIVATE AND ENABLE

The first and most important choice a leader makes is the choice to serve, without which one's capacity to lead is severely limited.

Robert Greenleaf, founder of the modern servant leadership movement

Servant Leadership

Servant leaders give more of themselves, not because they have to but because they want to. Servant leaders are transparent, honest, and even vulnerable. Servant leadership has 4 Principles.

Principle 1: Encourage Diversity of Thought

Great leaders assemble a diverse group of people from all walks of life. People want to be a part of a diverse team that fosters a welcoming environment. Servant leadership encourages everyone to think outside the box, and considers every perspective when tasked with moving forward. The final decision is the byproduct of collaboration and the exchange of ideas.

Diversity and inclusion are a business priority. Companies with great diversity outperform their peers by a significant margin.[104] Diverse organizations are more likely to develop innovative leaders. Progress and innovation depend more on diverse groups working together and capitalizing on their individuality than on lone thinkers with high intelligence. Sustainable business

104 Josh Bersin. 2015 (updated 2019). Why Diversity and Inclusion Has Become a Business Priority. https://joshbersin.com/2015/12/why-diversity-and-inclusion-will-be-a-top-priority-for-2016/

performance today depends on your ability to attract and engage a highly diverse workforce.

Principle 2: Create a Culture of Trust

Trust is one of the hardest things to regain once it is broken. As you learned in the chapter on trust, you create a culture of trust by behaving with integrity, exhibiting competence, and being a team player. Also, you must clearly communicate to everyone the organization's vision, mission, strategy, and expected organizational and individual values.

Principle 3: Have an Unselfish Mindset

You must understand that it is not about you but the others who do much of the work. Results, whether impact or profits, happen because of people. Great leaders help facilitate the success of others, make everyone feel valued, and recognize other's contributions to the overall success of the organization.

Principle 4: Foster Leadership

Leaders intent on building a great team understand the need to develop the next generation of leaders. Fostering leadership comes in many forms, including coaching, mentoring, and growth. Leaders make time to teach, offer words of encouragement, and answer questions. Great leaders give back.

Extrinsic and Intrinsic Motivation

Motivation is the desire to act in service of a goal and to attain meaning and purpose. Motivation drives human behavior. Without motivation, people can suffer mental illness or depression.

One framework used for understanding motivation is Maslow's hierarchy of needs.[105] According to Maslow, an American psychologist, humans are inherently motivated to better themselves and move toward expressing their full potential.

105 A.H. Maslow. 1943. A theory of human motivation. Psychological Review (50):370–396

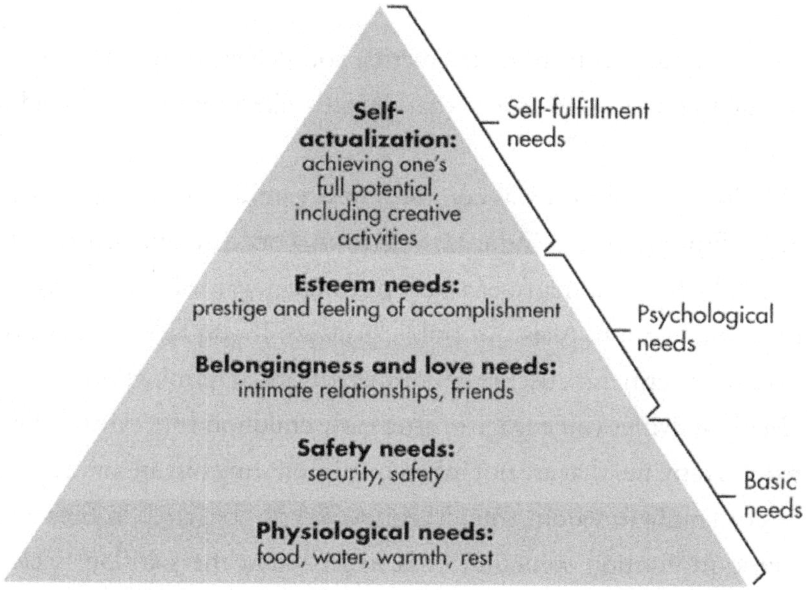

Source: Maslow 1943.

Our initial motivation is for basic needs – food, water, warmth, and rest. Once our basic needs are met, we are motivated to attain security and safety. Secure and safe, we seek belongingness, and love. Then we seek self-esteem and self-fulfillment. At our pinnacle, we transcend ourselves and help others meet their needs.

There are 2 fundamental types of motivation: 1) extrinsic and 2) intrinsic. You should be clear which type of motivation you rely on when leading others.

Extrinsic Motivation

Do you agree or disagree with the following statements?

- Employees fear taking responsibility.
- People desperately need direction.
- People crave security.
- People dislike work and avoid it when possible.

- People must be coerced and controlled.

If you agreed with these statements, you believe people are extrinsically motivated and respond to external rewards like money, fame, grades, and praise.[106]

Extrinsic motivation was common in organizations throughout the last century. Employees were induced to perform work through carrots and sticks. If you did what your manager asked, you were rewarded with a carrot - monetary compensation, praise, or public acclaim. If you failed to perform you faced a stick - criticism, decreased compensation, or termination.

Many activities you engage in after early childhood are extrinsically motivated.[107] Activities that are not inherently interesting but are valued by supervisors or family to whom you feel connected or obligated. In the workplace extrinsic motivation sometimes has merit when the solution is clear and straightforward, and compensation is fair and adequate.[108]

Extrinsic motivation is ineffective when the solution is unclear and complicated, and compensation is unfair and inadequate. The power of compensation to motivate performance lasts for a short time, perhaps as little as a week.[109] People typically work only enough to trigger the reward but no more.

Studies reveal a very weak association between salary and job satisfaction.[110,111] In fact, extrinsic awards can demotivate employees especially when tasks are interesting or enjoyable rather than boring or meaningless. [112]

106 Burrhus Skinner. 1953. Science and Human Behavior. Macmillan. New York, New York.

107 Richard Ryan and Edward Deci. 2000. Intrinsic and extrinsic motivations: Classic definitions and new directions. Contemporary Educational Psychology. 25:54–57.

108 Judy Cameron, Katherine Banko, and David Price. 2001. Pervasive negative effects of rewards on intrinsic motivation: The myth continues. The Behavior Analyst. 24:1–44.

109 Doug and Polly White. 2015. Money Is Nice, But It's Not Enough to Motivate Employees. June 23. https://www.entrepreneur.com/article/247333.

110 Timothy Judge, Ronald Piccolo, Nathan Podsakoff, John Shaw, and Bruce Rich. 2010. The relationship between pay and job satisfaction: A meta-analysis of the literature. Journal of Vocational Behavior 77:157–167.

111 Nikki Blacksmith and Jim Harter. 2011. 28 October. Workers not engaged in their jobs. https://news.gallup.com/poll/150383/majority-american-workers-not-engaged-jobs.aspx.

112 Edward Deci, Richarch Koestner, and Richard Ryan. 1999. A meta-analytic review of experiments examining the effects of extrinsic rewards on intrinsic motivation.

Extrinsic motivation can also diminish initiative and learning, especially when learning is complex or requires creativity.[113,114]

Intrinsic Motivation

Do you agree or disagree with these statements?

- Creativity and ingenuity are widely distributed in the population.
- Interest in work is as natural as play.
- Motivation is made, not born.
- Under the right conditions, people will accept and even seek responsibility.

If you agree, you believe in intrinsic motivation that originates inside a person and reflects satisfaction of psychological needs like interest or joy. Acting in self-interest results in high-quality learning, creativity, knowledge, and skill development.[115,116]

Intrinsic motivation is necessary for heuristic work that requires experimenting to solve novel problems. Employees who are intrinsically motivated are three times more engaged, more satisfied, and less likely to resign than extrinsically motivated employees.[117]

113 C. Benware and E. Deci. 1984. Quality of learning with an active versus passive motivational set. American Educational Research Journal. 21:755–765.

114 W.S. Grolnick and R.M. Ryan. 1987. Autonomy in children's learning: An experimental and individual difference investigation. Journal of Personality and Social Psychology. 52:890–898.

115 Richard Ryan and Edward Deci. 2000. Intrinsic and extrinsic motivations: Classic definitions and new directions. Contemporary Educational Psychology. 25:54–57.

116 Teresa Amabile. 2013. Componential theory of creativity. Encyclopedia of Management Theory (Ed. Eric Kessler). Sage Publications.

117 Yoon Jik Cho and James Perry. 2011. Intrinsic motivation and employee attitudes: Role of managerial trustworthiness, goal directedness, and extrinsic reward expectancy. Review of Public Personnel Administration. 32:382–406.

Self-determination Theory

Self-determination theory is a concept of motivation and personality that addresses three universal and innate psychological needs - competence, autonomy, and psychological relatedness.[118,119,120]

The theory distinguishes between extrinsic and intrinsic motivation with the quality of the motivation experience and performance higher with intrinsic motivation. Tangible extrinsic rewards can undermine intrinsic motivation.[121]

Having earlier learned how to build trust and create a shared vision, you can use self-determination theory to catalyze intrinsic motivation and enable people to succeed and be productive.[122,123,124]

Competence

Competence is the universal desire to control the outcome and experience mastery through accomplishment. The feeling of effectiveness at meeting challenges and opportunities, and to demonstrate and grow skills.

As a leader, you can enable competence in others.

- Encourage and support learning.
- Demonstrate an enthusiastic and genuine belief in the capacity of others.

118 Edward Deci and Richard Ryan. 2012. Self-determination theory. In P. A. M. Van Lange, A. W. Kruglanski, & E. T. Higgins (Eds.), Handbook of Theories of Social Psychology (pp. 416–436). Sage Publications Ltd. Thousand Oaks, California.

119 Marylene Gagné and Edward Deci. 2005. Self-determination theory and work motivation. Journal of Organizational Behavior 26(4), 331–362.

120 Richard Ryan and Edward Deci. 2000. Self-determination theory and the facilitation of intrinsic motivation, social development, and well-being. American Psychologist 55(1):68–78.

121 Edward Deci. 1971. Effects of externally mediated rewards on intrinsic motivation. Journal of Personality and Social Psychology. 18:105–115.

122 Susan Fowler. 2014. What Maslow's hierarch won't tell you about motivation. Harvard Business Review. November. https://hbr.org/2014/11/what-maslows-hierarchy-wont-tell-you-about-motivation

123 Teresa Amabile and Steven Kramer. 2011. The Progress Principle. Using Small Wins to Ignite Joy, Engagement, and Creativity at Work. Harvard Business Review Press. Boston, MA.

124 James Kouzes and Barry Posner. 2017. The Leadership Challenge: How to Make Extraordinary Things Happen in Organizations. Wiley and Sons, Hoboken, New Jersey.

- Let people stretch their boundaries.
- Create and use a training budget.
- Include a personal development plan in everyone's annual performance objectives.
- Acknowledge what people do well.
- Make yourself available to discuss challenges and opportunities
- Talk with staff instead of talk at staff.
- Treat mistakes as learning opportunities.

Autonomy

Autonomy is the universal urge to be in control of your life and aligned with your purpose. To do something valuable and meaningful. You will stop trying if your actions do not affect what happens to you. Why expend energy when the results are beyond your control?

As a leader, you can enable autonomy.

- Delegate authority and let people lead themselves.
- Work with staff to reach a common understanding of the desired outcome and a framework for how to achieve the outcome.
- Allow people to make choices consistent with the outcome and within the bounds of the framework.
- Agree on a set of benchmarks to ensure progress.
- Provide timely and constructive feedback.
- Avoid applying pressure to perform.
- Avoid the promise of a cash incentive.

Psychological Relatedness

Relatedness is the universal need to interact with and to be connected to others. It is caring for others and having others care for you. The desire to be part of something bigger than yourself. Neurological research illustrates that our most productive, innovative, and collaborative times at work happen when we feel like we are a part of the team.[125]

As a leader, you can enable relatedness.

- Administer policies and practices equitably.
- Talk to people often enough to maintain a relationship.
- Ask staff if they feel valued, and if not, what would make them feel valued.
- Ask staff if they are open to suggestions.
- Encourage face-to-face meetings.
- Help people understand how they fit into the organization.
- Facilitate relationships among people to enhance team effectiveness.
- Develop cooperative goals and roles.
- Reward reciprocity.
- Share knowledge.
- Involve staff in decision-making.
- Be transparent about why the organization makes its decisions.
- Help others understand how the organization operates.

Mindfulness

Mindfulness, the practice of intentionally paying attention in the present moment without judgment, is a factor in motivation theory.[126,127] Mindfulness is positively associated with intrinsic motivation and negatively associated with extrinsic motivation. Mindfulness induces well-being, work

125 Josh Bersin. 2015 (updated 2019). Why Diversity and Inclusion Has Become a Business Priority. https://joshbersin.com/2015/12/why-diversity-and-inclusion-will-be-a-top-priority-for-2016/

126 Richard Ryan and Edward Deci. 2017. Self-determination theory: Basic psychological needs in motivation, development, and wellness. New York, NY: Guilford Publications.

127 James Donald et al. 2019. Mindfulness and its association with varied types of motivation: A systematic review and meta-analysis using self-determination theory. Personality and Social Psychology Bulletin. November.

engagement, performance, memory function, and reduces stress, anxiety, and substance abuse.[128,129,130,131]

Mindful individuals are less likely to be motivated by extrinsic financial rewards, social recognition, guilt, or shame.[132] Controlling environments may demotivate mindful individuals, reducing engagement and performance and increasing stress, anxiety, and substance abuse.

Assessing Your Ability to Motivate and Enable

As a leader, you want to facilitate intrinsic motivation and avoid or minimize extrinsic motivation. You strive to support competence, allow self-governance, and encourage relationships and mindfulness.

As a final check on your readiness to motivate and enable others, read the following statements, and consider which actions disable others and which actions enable others.

1. I didn't have any input.
2. I was given responsibility but no authority.
3. Information essential to my work was withheld.
4. My supervisor argued with me in front of colleagues.
5. My supervisor did not support my decisions even after telling me they would back me up.
6. No one listened.
7. People ignored me.
8. People picked me apart.

128 Alison Christie, Paul Atkins, and James Donald. 2017. The meaning and doing of mindfulness: The role of values in the link between mindfulness and well-being. Mindfulness 8(2), 368–378.

129 Jochen Reb, Jayanth Narayanan, and Sankalp Chaturvedi. 2015. Leading mindfully: Two studies on the influence of supervisor trait mindfulness on employee well-being and performance. Mindfulness 6(1):111–122.

130 Kirk Brown and Richard Ryan. 2003. The benefits of being present: Mindfulness and its role in psychological well-being. 2003. Journal of Personality and Social Psychology. 84:822–848.

131 David Brown, Adrian Bravo, Corey Roos, and Matthew R. Pearson. 2015. Five Facets of Mindfulness and Psychological Health: Evaluating a Psychological Model of the Mechanisms of Mindfulness. Mindfulness 6(5): 1021–1032.

132 James Donald et al. 2019. Mindfulness and its association with varied types of motivation: A systematic review and meta-analysis using self-determination theory. Personality and Social Psychology Bulletin. November.

9. Someone took credit for my work.
10. All the financial data were shared with me.
11. I was able to exercise my discretion.
12. I was asked to take on an important project I had never done before.
13. My supervisor gave me the chance to learn new skills and the opportunities to apply them.
14. My supervisor took the time to let me know how I was doing.
15. My supervisor publicly expressed confidence.
16. My supervisor told others about the great job I was doing.
17. The organization invested resources in me to help me do the job.

The first nine statements represent actions that disable people, reduce their well-being and productivity, and lessen the potential of the organization. You demotivate and disable others when you do not let them provide input, give them responsibility but not authority, and withhold essential information.

You demotivate and disable others when you argue with them in front of their supervisor or colleagues, or fail to support their decisions after you promised to. You demotivate and disable others when you do not listen, or allow colleagues to ignore or criticize them. And perhaps worst of all, you demotivate and disable others when you allow someone else to take credit for their work.

The final eight statements represent actions that enable people, increasing their well-being and productivity and enhancing the potential of the organization. You motivate and enable others when you share financial information, allow them to exercise their discretion, take on an important project, and give them the chance to learn and apply new skills.

You motivate and enable others when you encourage their efforts, publicly express confidence in them, and tell others what a great job they are doing. And most of all, you motivate and enable others when the organization invests resources in them to help do the job.

To lead, you must embrace enabling actions and avoid disabling actions.

PART 4
SHARE LEADERSHIP

WHY SHARE LEADERSHIP

A leader is best when people barely know he exists, when his work is done, his aim fulfilled, they will say: we did it ourselves.

Lao Tzu, ancient Chinese philosopher and writer

There are several compelling reasons for sharing leadership. Sustaining healthy lands and waters is complicated. Solutions require extensive knowledge and input from multiple experts, acceptance of a new idea, and stakeholder willingness to change.

Technical Problems and Adaptive Challenges

Each of us has a profession characterized by technical problems or adaptive challenges.[133] Technical problems:

- Are easy to identify.
- Have known solutions.
- Can be solved by a single expert or within an existing structure.
- Have solutions generally accepted by stakeholders.
- Have solutions quickly implemented, often by edict.

Technical problems are addressed by putting in place known solutions. All you need is a toolbox and the training to use the right tool. For example, removing an invasive plant requires pulling, cutting, or herbicide application.

133 Ronald Heifetz, Alexander Grashow, Marty Linsky. 2009. The Practice of Adaptive Leadership. Harvard Business Review Press, Boston, Massachusetts, U.S.

Once you know the solution, you only need to choose the correct method each time you face the problem.

In contrast, adaptive challenges:

- Are difficult to identify.
- Require changes in values, beliefs, roles, relationships, and approaches.
- Are best solved by people who own the problem.
- Require solutions developed by multiple experts working together.
- Require solutions based on experiments and new discoveries.
- Require engaging stakeholders.
- Have solutions slow to implement.

Keeping lands and waters healthy are adaptive challenges. For example, permanently eradicating an invasive plant requires you to understand the ecological elements and processes that sustain the land, determine which stressors have degraded the land and will impact the land in the future, develop and implement a strategy to exclude or minimize stressors, and monitor results and manage adaptively.

Socio-economically, you need to know the ecosystem service values of the healthy and impaired land and water and the cost for offsetting lost services. Socio-politically, you need to understand the current and potential policies for conserving ecosystem services, parks, recreation, housing, jobs, transportation, safety, water, and waste treatment.

Burden of Knowledge

Another reason for sharing leadership is the burden of knowledge. As an individual committed to solving the challenge of declining lands and waters, you must accumulate knowledge by learning more or by narrowing your expertise. However, choosing to learn more leaves less time for work and narrowing your expertise reduces your capability to identify and implement solutions.

In a study of United States patents, the exclusive right granted by the government to an inventor to manufacture, use or sell an invention, the age of first innovation was trending upward at 0.6 years per decade[134].

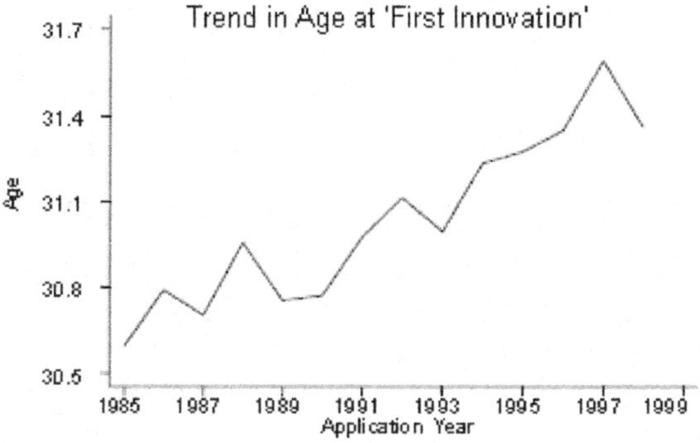

Source: Benjamin Jones. 2005.

And the size of the team applying for patents was increasing by 17 percent per decade.

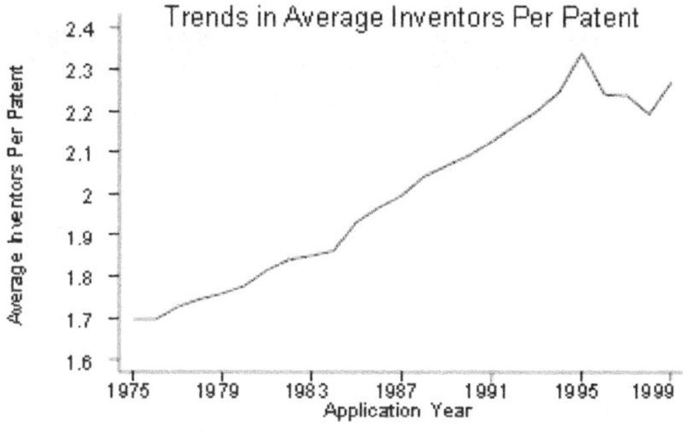

Source: Benjamin Jones. 2005.

134 Benjamin Jones. 2005. The Burden of Knowledge and the Death of the Renaissance Man: Is Innovation Getting Harder? National Bureau of Economic Research, Cambridge, Massachusetts, U.S.

The first figure illustrates that adaptive challenges are taking increased learning. The second figure suggests a solution – working with others.

Collective Intelligence

In fact, working with others is a solution to addressing increasingly complex challenges. Studies at the Massachusetts Institute of Technology identified a group general collective intelligence factor akin to an individual's general cognitive intelligence factor.[135,136] Group intelligence is a better predictor of group performance on multiple tasks than average or maximum individual intelligence. Groups are smarter than individuals.

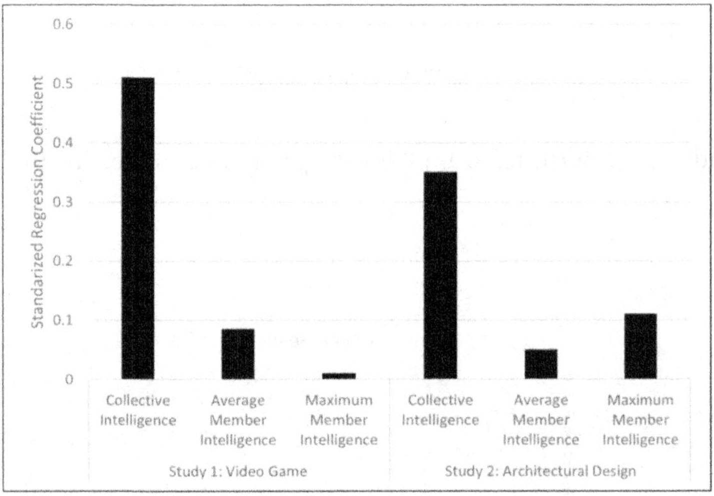

Source: Woolley et al. 2010.

Collective intelligence is most effective when group members:

- Share a common purpose.

135 Anita Woolley, Christopher Chabris, Alexander Pentland, Nada Hashmi, and Thomas Malone. 2010. Evidence for a Collective Intelligence Factor in the Performance of Human Groups. Science. Vol. 330, 29 October 2010, pp 686–688.

136 Pentland, Alex. 2012. The New Science of Building Great Teams. Harvard Business Review, http://hbr.org/2012/04/the-new-science-of-building-great-teams/ar/pr.

- Recognize tangible benefits e.g., funding, work products, and extended resources.
- The group supports communication and multiple, dynamic hubs.
- Group leadership manages group dynamics.
- Clients and stakeholders highly regard the group.
- Group members communicate frequently.
- Group members talk and listen in equal measure.
- Group members engage in frequent informal communication.
- Group members explore ideas and information outside the group.

Distributed groups and groups with multiple languages can be less effective than localized, same language groups.

Ownership

Another reason to share leadership is to convey ownership.

Committing to a course of action requires acceptance of an idea and willingness to change. In the decision-making and change chapters, you learned that you rely extensively on your limbic system. This is a system that acts based on what it knows, and is reluctant to change without discovering for itself.

How responsive are you when someone tells you they have a solution to a problem, and you will have to do something you do not understand or agree with? Contrast that with someone asking you to help them find a solution to a problem. In which situation are you more likely to support the solution? For most people, being part of the solution facilitates acceptance of an idea and willingness to change.

Ownership is conveyed by identifying and meaningfully engaging your stakeholders. Stakeholders are those organizations and individuals that can help you find a solution and implement it. Ownership does not mean everyone gets what they want, but that everyone had a chance to be heard and to understand the *why* of a solution. Without stakeholder ownership your chance of success is reduced.

Networks

Networks provide a useful context for working with others to overcome knowledge burdens, benefit from collective intelligence, and convey ownership. Networks consist of nodes (people, organizations, ideas), and links (bonds and bridges) that connect the nodes.[137] A cluster is a group of nodes that are connected more to each other than to other nodes outside the cluster. Bonds are links within a cluster, and bridges are links between clusters.

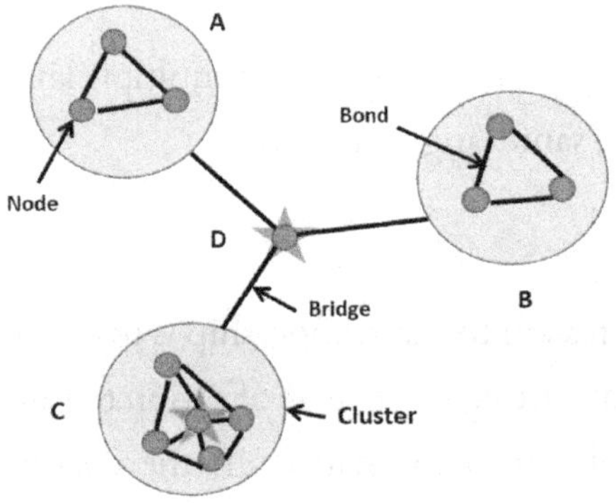

Source: Adapted from Bruce Hoppe and Claire Reinelt. 2010.

In the figure, the star in cluster C indicates that the nodes are highly connected. Highly connected clusters have a high degree of affinity, trust, support, sense of community, and efficiency. The star in node D indicates that the organization's role is as a bridge between clusters. Bridging creates access to the latest ideas, innovations, and resources.

Leading with a Network Mindset

Networks are an elegant solution to shared leadership but working with others is much more than sharing data or techniques. Working with others has costs, especially in the short term. There are investments of time,

137 Bruce Hoppe and Claire Reinelt. 2010. Social Network Analysis and the Evaluation of Leadership Networks. The Leadership Quarterly, 21:600–619.

resources, and energy in building and maintaining relationships. No one organization has control over outcomes and there are coordination and communication challenges.

Historically, leadership promoted a heroic, individually centered approach. The emphasis was on improving individual performance. By contrast, working with others requires cultivating collective leadership. Contributing to a group over which you have little or no authority requires a different mindset, one that is more distributed, relational, and interdependent. Leaders working in groups practice openness and transparency, let go of control, and enable the leadership potential of everyone.

Leaders in a network or other group setting connect and weave. Intentionally introducing and linking people together, they strengthen bonds and build bridges to expand the group's reach, influence, and innovation. They enable self-organizing, establishing a common purpose and creating opportunities for individuals and groups to act on their ideas.

Leaders in networks embrace risk-taking. They commit to making risk safe and to enabling continuous learning, both of which are essential for adapting to continuous change and increasingly complex challenges.

Leadership in networks arises from the capacity of groups to coordinate their actions and to learn together for greater impact. Here are four strategies designed to develop leadership in the context of shared work.

1. *Convene and build relationships across boundaries.* Host meetings that enable dialogue among people with different perspectives to build trust and connect ideas and experiences.
2. *Introduce a group mindset into a leadership curriculum.* Challenge people to question their assumptions about leadership, to be authentic, and serve others. Conduct and share social network analyses to reveal relationship patterns. Introduce basic collaboration tools such as blogging, wikis, and Twitter to facilitate communication.
3. *Facilitate action learning.* Create opportunities for experimentation, risk-taking, and accountability based on actual challenges.

Opportunities that help groups clarify their purpose, identify, and map desired results, and decide on actions that increase their ability to achieve the desired results.

4. *Invest in communities of practice.* Communities of learning and practice provide moral support, peer assistance, tools, techniques and collaborative problem-solving, especially to groups adopting new behaviors and growing their knowledge.

The Benefits of Shared Leadership

How can shared leadership and working with others help you sustain healthy lands and waters?

- *Better outcomes*: Working together brings more experiences to a challenge, increasing the chances of resolution.
- *Efficient learning*: Ready access to the ideas and experience of others.
- *Less work for any individual*: Work is shared among the group without compromising results.
- *Less risk*: Working with others encourages shared responsibility for outcomes and enables peer review.
- *Agility*: Groups are better able than individuals to develop and implement ideas that arise.
- *Innovation*: Working with others promotes creativity through the integration of different ideas and perspectives.
- *Improved morale*: Working with others allows group members to take greater responsibility for decision-making.
- *Mutual support*: Working with others encourages people to achieve goals they might not reach on their own
- *Funding*: Working with others gives funding sources more confidence that the project can address complex challenges.

- *Fun*: Working with others is energetic, and shared success promotes a greater sense of accomplishment than individual success
- *More durable decisions*: Working with others facilitates decisions that have broad support and are less susceptible to opposition.
- *Early adopters*: Your members are usually the first to adopt the group's solutions.
- *Future*: Today's partners will think of you tomorrow when they face a new challenge.

COMMUNICATE EFFECTIVELY

Who you are is speaking so loudly that I cannot hear what you are saying.

Ralph Waldo Emerson, American essayist, lecturer, philosopher, and poet

Sometimes your urgency, movement, tone, and volume overshadow what you are trying to say. Communication is an exchange of thoughts, opinions, and information. Effective communication enables understanding between individuals and among groups. Meaningful interaction can bring people closer together, bridge gaps, and promote learning. But only if you can communicate effectively.

Strong communication skills are essential for leadership. Effective communication enables you to develop strategy, manage employees, transmit expectations, assign tasks, and persuade others to support the organization or an initiative.

Effectively engaging employees has a substantial positive impact your organization's profitability, productivity, employee turnover, safety incidents, organizational downsizing, absenteeism, and product and service quality.[138] Additional research is finding links between employee engagement and health and well-being.

138 James Harter, Frank Schmidt, Sangeeta Agrawal, Stephanie Plowman and Anthony Blue. 2016. The Relationship Between Engagement at Work and Organizational Outcomes. Q12® Meta-Analysis: Ninth Edition. Gallup, Washington, D.C.

Effective communication enables you to skillfully address emotionally and politically risky issues. This skill predicts with nearly 90 percent accuracy which projects will fail months or years in advance.[139]

Communication serves several purposes.

- To share knowledge without bias.
- To express feelings about yourself, others, policies, events, places, and objects.
- To create and maintain your social standing.
- To stimulate your and others' imaginations.
- To persuade people to adopt a set of beliefs or to act.
- In a group, communication enables the flow of information, commitment to a common purpose, and coordinated action.

Leaders have a responsibility to communicate effectively. You have information that can make lands and waters healthy. Or you know how to obtain that information. You have an obligation to share what you know with others. And you should listen to what others can share with you.

When you communicate, whether talking or listening, focus your communication on solutions, not problems. Learn how to communicate more effectively by seeking and accepting criticism. And communicate authentically, passionately, and persuasively.

Effective communication is a complex topic touching on understanding your listeners, clarity, storytelling, body language, and persuasion.

Understand Mindset

Both you and the person you are communicating with must commit to understanding the information being communicated. To understand, you must commit to understanding mindset - the established set of attitudes you each hold. Your brain understands meaning in relation to

139 Kerry Patterson, Joseph Grenny, Ron McMillan, and Al Switzler. 2011. Crucial Conversations: Tools for Talking When Stakes are High. McGraw-Hill.

other meanings it experienced rather than as new, discrete entries.[140] How you interpret information is dependent on previous experiences.

For example, someone with a growth mindset:

- Believes they can learn anything they want.
- Perseveres when frustrated.
- Wants to challenge themselves.
- Believes they learn from failure.
- Is inspired by others' success.
- Believes that their effort and attitude determine everything.

In contrast, someone with a fixed mindset:

- Believes they are either good at something or they are not.
- Gives up when frustrated.
- Does not like being challenged.
- Believes they are no good when they fail.
- Needs someone to tell them they are smart.
- Feels threatened by others' success.
- Believes that their abilities determine everything.

To communicate effectively, a growth mindset person and a fixed mindset person must understand their different perspectives, engage in a process that lets each share and listen, and then try to find common ground for moving forward. You can review the change personalities in the chapter on promoting change for more insight.

Culture also affects your capacity for communication. Culture imparts a mindset formed over years from exposure to family, friends, school, and society. Your mindset makes it easier to understand compatible information and harder to understand incompatible information.

140 Ben Thomas. 2012. Meaning on the brain: How your brain organizes reality. https://blogs.scientificamerican.com/guest-blog/meaning-on-the-brain-how-your-mind-organizes-reality/. 26 December.

Know Your Audience

You must know your audience, and they you, to achieve effective communication. Start by asking what your audience expects from you, and compare that with what you expect from them.

For example, does the audience want to share your knowledge, hear your opinion about a topic, participate in a social event, be stimulated by new ideas, or be persuaded to adopt a position? Are your expectations to entertain, inform, or persuade? Can you craft a communication that satisfies both sets of expectations? If not, you risk wasting everyone's time and making a mistake that causes you embarrassment and lost credibility.

How do you learn about your audience? The most direct way is to speak to them or a representative subset. If not possible, you can speak to someone familiar with the audience.

If neither is possible, assess their characteristics to the best of your abilities. What is their age, gender, culture, status, education, and familiarity with your topic? If you are trying to be persuasive, is the audience likely to be swayed by logic or emotion? Do they have a preferred communication style, i.e., face-to-face, video, telephone, or written? Social media can help you understand a small audience.

Not everyone believes you should consider your audience. The distinguished William Zinsser, author of *On Writing Well,* believes you should write for yourself.[141] He maintains that every reader is different, and most readers do not know what they want until they read it. I agree to a point. Your communication, whether written or verbal, should be authentic and reflect your personality. But authenticity should be shaped to an audience and aligned with your purpose for communicating.

141 William Zinsser. 2001. On Writing Well: The Classic Guide to Writing Nonfiction. Quill: A Harper Resource Book. New York, New York.

Listen Actively

Listening is the most fundamental component of interpersonal communication skills. Yet, listening is an abstract and less effective form of communication than observing or doing.[142]

Listening effectiveness can be improved through *active listening.*[143] Clinical and research evidence show that active listening produces individual personality change and group development. You become more emotionally mature, more receptive to new experiences, less defensive, more democratic, and less authoritarian.

When you listen to others, they listen to themselves more carefully and become better at communicating what they are feeling and thinking. Group members tend to listen more, be less argumentative, and more open to accepting other points of view.

Carl Rogers and Richard Farson, psychologists and educators, developed active listening in the 1950s as a therapeutic technique to promote positive change in clients.[144] Today, active listening is a communication technique used in counselling, training, and conflict resolution.

The technique requires the listener to repeat back what they hear to the speaker by way of re-stating or paraphrasing in their own words. This confirms that what the listener heard is what the speaker meant, enhancing understanding. Fundamentally, the technique avoids challenging the speaker's self-image and instead helps them explore what they are thinking and feeling.

Active listening reflects a conscious effort to hear and understand using all senses. Elements of active listening include the following.[145]

- *Listening for Total Meaning.* The speaker is communicating content and feelings, which combined, transmit meaning. Meaning is what

142 Edgar Dale. 1969. Audiovisual Methods in Teaching. Holt, Rinehart & Winston, Dryden Press. New York, New York.

143 Carl Rogers and Richard Farson. 1987. Active Listening. Excerpt from Communicating in Business Today (R. Newman, M. Danzinger, and M. Cohen, Eds.). D.C. Heath & Company.

144 Carl Rogers and Richard Farson. 1957. Active Listening. Martino Publishing. Mansfield Centre, Connecticut.

145 Carl Rogers and Richard Farson. 1987. Active Listening. Excerpt from Communicating in Business Today (R. Newman, M. Danzinger, and M. Cohen, Eds.). D.C. Heath & Company.

you strive to understand. Maintain eye contact, pay attention to the speaker's body language, avoid environmental distractions, and suppress the urge to respond.

- *Respond to Feelings.* Feelings are the key to understanding the speaker. Try to decipher what the speaker is trying to communicate, what they are feeling, and how they view the situation or issue. To encourage feelings, nod, smile, maintain an open and interested posture - lean slightly forward, tilt your head, allow yourself to automatically reflect the speaker's expressions, and avoid crossing your arms. Encourage the speaker to continue.
- *Note All Cues.* Non-verbal communication is important for determining meaning. Does the speaker hesitate, change his or her pitch or tone, speak loud and clear, or mumble? What do the speaker's facial expressions, body posture, eye movements, and breathing communicate?
- *Reflect in Your Own Words.* Your own experiences can distort what you hear. When the speaker invites feedback or has finished talking, paraphrase what they said and ask them to confirm that this is what they meant. This can take the form of statements or clarifying questions. Continue to paraphrase until the speaker agrees you have a shared understanding of the situation or issue. Only then should you offer your own thoughts.

Active listening is a powerful tool. For it to work you must be sincere in wanting to understand the speaker's meaning and be open to changing yourself. Practice active listening when speaking with others. Listen for total meaning, respond to emotions, note all cues, and reflect in your own words. As your skill improves, notice that you are more receptive and less defensive, and that those you speak with are better communicators.

Be Clear

Clearly communicating content and meaning happens at several levels. You should start by thinking about what you are trying to communicate, and why it is relevant to your reader or listener. Organize your material, simplify, and use good grammar.

The Message Box

Nancy Baron has developed an elegant tool - the message box - to organize your information to address the few key messages your audience needs.[146]

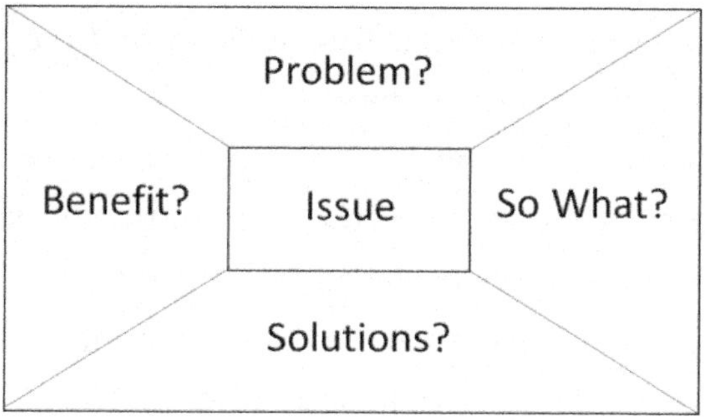

Source: Nancy Baron. 2010.

At the heart of the message box is the issue or topic you want to communicate. For example, the health of our lands and waters is diminishing. The issue can be framed as a fact, opinion, question, or conclusion depending on whether your purpose is to share knowledge, express a feeling, establish social standing, stimulate imagination, or persuade.

Why is this issue a problem? Continuing the healthy land's theme, you might describe the problem as unsustainable land and water use practices.

So what? speaks to why the audience should care that the condition of our lands and waters is diminishing and our protection efforts are inadequate. Here you might communicate the ecosystem values of healthy lands and

146 Nancy Baron. 2010. *Escape From the Ivory Tower: A Guide to Making Your Science Matter*. Island Press. Washington, D.C.

waters, or conversely the death and destruction that accompany unsustainable practices.

Solutions speak to potential actions, for example, augmenting regulations, preserving more land, educating the public, sharing knowledge, and advancing practices.

Lastly, the benefits of the solutions are explained. In the healthy lands and waters context the ultimate benefits are human well-being and prosperity.

The message box is designed to help you prepare for effective communication. The box does not imply that you should communicate every part of the box in every situation. Or that you need to adhere to the *issue, problem, so what, solutions, benefit* sequence. Your audience and purpose should guide how you communicate your key messages.

You can practice clarifying your message by choosing an issue or topic, preferably one related to your purpose, and constructing a message box. Define your audience's expectations and receptivity to emotion and logic, the goal of your communication (i.e., entertain, inform, or persuade), and the communication mode (i.e., face-to-face, video, telephone, or written). Write clear statements about the problem, why it matters, a solution, and the benefits of the solution.

Simplify

The next step to achieving clarity in your writing and speaking is to simplify.[147,148,149] Limit yourself to one or a few key messages. Choose and maintain a pronoun (e.g., I, you, he, she), tense (e.g., past, present, or future), and tone (e.g., casual, formal).

Use active voice and verbs (in the active voice, the subject [noun] of a sentence performs an action [verb] on an object, e.g., Pat conserved the land).

147 Nancy Baron. 2010. Escape From the Ivory Tower: A Guide to Making Your Science Matter. Island Press. Washington, D.C.

148 Glenn Leibowitz. 2016. All good writers do this 1 thing to significantly improve their writing. https://www.inc.com/glenn-leibowitz/do-this-1-thing-to-significantly-improve-your-writing.html. 25 May.

149 William Zinsser. 2001. On Writing Well: The Classic Guide to Writing Nonfiction. Quill: A Harper Resource Book. New York, New York.

Vary sentence length and provide context by comparison to familiar situations or objects using magnitude (size or extent) instead of statistical significance.

If you are writing, read the text out loud to yourself to see how it sounds and how comfortable you are speaking it. Avoid clichés, phrases or opinions that are overused and betray a lack of original thought.

Avoid jargon, special words or expressions that are used by a particular profession or group and are difficult for others to understand. For example, *Biodiversity* is jargon for biological diversity, which means the variety of life on the planet and how it interacts within habitats and ecosystems. Biological diversity, itself, requires explanation for many audiences.

Examine every word to determine if it serves a purpose. Avoid long, obscure words if a short, common word will do. For example, serendipity means the occurrence and development of events by chance in a happy or beneficial way. Chance or accident are clearer.

Do not ask your audiences to do math. Use frequencies (the rate at which something occurs or is repeated over a particular period or in each sample) instead of probabilities (the likelihood of something happening or being the case).

Grammar

Grammar also contributes to clear and effective written or verbal communication. Adherence to a few principles will facilitate reader or listener comprehension.[150]

The subject (noun: a person or thing that is being discussed, described, or dealt with) should be followed by its verb (expresses an act, occurrence, or mode of being).

The topic position at the beginning of a sentence should provide linkage between preceding information and upcoming information. The subject should also appear in the topic position.

150 George Gopen and Judith Swan. 2018. The science of scientific writing. American Scientist. https://www.americanscientist.org/blog/the-long-view/the-science-of-scientific-writing.

Sentences should generally be limited to one point of emphasis and emphatic material (that which you deem most worthy of attention) should be placed in the stress position at the end of a sentence.

Tell a Story

Storytelling is perhaps the most effective way to communicate. The power of storytelling derives from how your brain functions.

During verbal communication, your brain and that of the person you are talking to couple to share information.[151] The degree of coupling depends on the quality of what you say. Backward sounds are registered by the part of the brain that processes sound. Scrambled words begin to induce alignment in the early language part of our brains. Sentences are registered by all language areas of the brain.

Only stories induce deep and broad alignment throughout the brain, including higher-order areas of the cortex responsible for thinking, perceiving, producing, and understanding language. Stories make you feel. Stories stick in your memory.

Reading stories also activates many areas of your brain.[152] Stories create a sensation of not just reading about the action, but experiencing the sensations it's describing.[153] This essential capacity helps you work with others and supports the relationships that form communities and broader society.[154,155]

Brain activation enables story readers to identify beliefs, emotions, and motivations. Stories that reflect your own experiences or a sense of who you

151 Uri Hasson. 2017. This is your brain on communication. https://ideas.ted.com/this-is-your-brain-on-communication/. 27 January.

152 Raymond Mar. 2011. The neural bases of social cognition and story comprehension. Annual Review of Psychology. 62:103–134.

153 Sadie Trombetta. 2017. What Does Reading Do To Your Brain? These 5 Effects Are Pretty Astounding. 11 August. https://www.bustle.com/p/what-does-reading-do-to-your-brain-these-5-effects-are-pretty-astounding-74676.

154 Michael Tomasello, Malinda Carpenter, Josep Call, Tanya Behne, and Henrike Moll. 2005. Understanding and sharing intentions: the origins of cultural cognition. Behavioral Brain Science. 28:675–735.

155 Anne Watson, Charisse Linkie-Nixon, Amy Wilson, and Laura Capage. 1999. Social interaction skills and theory of mind in young children. Developmental Psychology. 35:386–391.

are or want to be receive the most attention.[156] Reading, more than listening, gives your brain time to stop, think, process, and imagine.

How do you tell verbal or written stories? Stories typically have three to seven parts, with five balancing simplicity and complexity.[157,158,159]

Introduction

Your stories begin by preparing the audience for what is to come. You must answer who, what, why, when, and where to acquaint your audience with the characters, their thinking, the location, and the time. Provide enough background for your audience to bond with the main character, which could be you.

Conflict

Introduce a conflict or obstacle. The conflict may be internal (e.g., the character struggles with their conscience), with another character, or be an external obstacle (e.g., climate change, floods, drought, starvation). To create tension, the conflict or obstacle should be unavoidable.

Climax

Your climax is the turning point of the story. Often the most exciting and intense part of the story, your character faces the conflict or obstacle. Show your audience how they overcome or fail. In the context of healthy lands and waters, I recommend failures as subplots that ultimately are overcome providing a successful resolution. Keep hope alive.

156 Christin Scholz, Elisa Baek, Matthew Brook O'Donnell, Hyun Suk Kim, Joseph N. Cappella, and Emily B. Falk. 2017. A Neural Model of Valuation and Information Virality. Proceedings of the National Academy of Sciences. 114 (11) 2881–2886.

157 Kim Kautzer. 2016. 5 stages of storytelling: Help kids write more exciting narratives. https://writeshop.com/5-stages-of-storytelling.

158 Kristine Tucker. Accessed 27 November 2019. 5 Parts of a Plot in a Story. https://penandthepad.com/5-parts-plot-story-8394628.html.

159 Laura Carter. 2016. A 5-Part Story Structure For Beginners. 21 October. https://writerswrite.co.za/a-5-part-story-structure-for-beginners/.

Closure

Tie up loose ends after your character faces their conflict or obstacle. Address unresolved and outstanding tensions that are waiting for closure.

Conclusion

Show your audience how your character(s) has changed over time. Revisit who, what, why, when, and where so your audience has no unanswered questions.

Apply Fiction Techniques to Your Story

Whether speaking or writing, the techniques of fiction can be applied to nonfiction to give your story depth and resonance.[160] *The Lorax*, a children's book by Dr. Seuss, has sold more than 200 million copies and is revered for its whimsical characters and wonderful wordplay.

But the story is really a cautionary tale about respecting the environment and all living creatures that help us preserve the planet for ourselves and future generations. The message of *The Lorax* is so powerful that the book is widely used as a teaching aid for children's classes on environmentalism. And several university groups devoted to conservation have named themselves *The Lorax*.

Give your central character specific, relatable, physical, and psychological qualities, including something unique. Provide background so the audience can understand how life has shaped them. Evoke emotion in the audience through revealing the emotions of your character. And to express ideas, use anecdotes about your character.

Your story should have one idea and say something important.[161,162] Organize your story so it makes sense to your audience.[163] Be authentic, sharing yourself so the audience can connect with you whether you are the central character or the storyteller. Lead with something that induces the audience to

160 Sol Stein. 1995. Stein on Writing. St. Martin's Griffin. New York, New York.

161 Nilofer Merchant. 2013. Secrets from a TED2013 speaker: Preparing for the "Talk of one's life." 22 February. https://www.linkedin.com/pulse/20130222153141-1131485-secrets-from-a-ted2013-speaker-preparing-for-the-talk-of-one-s-life/.

162 Daniel Pink. 2016. 3 tips for TED speakers (and anyone else giving a talk). July 12. https://www.linkedin.com/pulse/3-tips-ted-speakers-anyone-else-giving-talk-daniel-pink/.

163 Nancy Baron. 2010. Escape From the Ivory Tower: A Guide to Making Your Science Matter. Island Press. Washington, D.C.

continue listening or reading, and end with an encapsulation of your one important idea.[164] Make sure your audience has a takeaway.

If your goal is healthy lands and waters, your story should inspire hope.[165] Hope motivates people to act. Hope assures people they can control the situation. Hope tells people they are valued and that the actions they take are important.

For speeches, know your story well enough that it sounds like a conversation and not a reading or memorization. Create an experience that is comfortable and engaging. And tailor your language, tone, cadence, and energy to match your purpose (i.e., share knowledge, express feelings, social standing, stimulate other's imaginations, or persuade).

Communicate Without Words

Voice and Gestures

Voice intonation, pace, and amplitude, and gestures, are also ways you communicate. How you use these signals can support or detract from your words.

For example, the success of telephone salespeople is accurately predicted by their ratio of speaking to listening, and fluctuation in voice amplitude and pitch.[166] Whether a marriage will work can be predicted by observing the pitch, rhythm, and tone of the couple's conversation.[167] How college students respond to their instructors is largely predicted by the instructor's body language.[168]

Longstanding research describes the ability of people to read one another with about 70 percent accuracy. Alex Pentland of the MIT Media Lab

164 William Zinsser. 2001. On Writing Well: The Classic Guide to Writing Nonfiction. Quill: A Harper Resource Book. New York, New York.

165 Ronald Swainsgood and James Sheppard. 2010. The culture of conservation biologists: Show me the hope! BioScience 60(8):626–630.

166 Mark Buchanan. 2007. The science of subtle signals. Strategy and Business Magazine. Issue 48. https://www.strategy-business.com/article/07307?gko=10cab.

167 Malcolm Gladwell. 2005. Blink! The Power of Thinking Without Thinking. Little, Brown, and Company. New York, New York.

168 Nalini Ambady and Robert Rosenthal. 1993. Half a minute: Predicting teacher evaluations from thin slices of nonverbal behavior and physical attractiveness. Journal of Personality and Social Psychology. 64:431–441.

attributes this ability to honest signals, an evolutionary model that predicts social species develop a reliable communication system to coordinate behavior between individuals.[169]

Pentland's group concentrates on four components of honest signals that seem to explain about 40 percent of communication. *Mimicry* is the reflexive copying of one person by another. *Activity* is movement that indicates interest and excitement. *Influence* is the extent to which one person matches the other's speech patterns. And *Consistency* is fluidity of speech, which is perceived as expertise.

Telephone salespeople that listen more than speak and fluctuate their voice signal interest. The pitch, rhythm, and tone of a married couple signals the warmth or coolness of the relationship. The body language of a favored instructor signals optimism, confidence, dominance, activity, enthusiasm, likeability, warmth, competence, and support.[170]

When you communicate, you tend to imitate facial expressions and gestures, and feel what others are feeling.[171,172] You have mirror neurons in the premotor and parietal cortex to thank or curse for this ability. Your mirror neurons respond when you see action and act.[173]

Conversely, the emotions you portray with your facial expressions and gestures directly influence others because of their mirror neurons. Facial expressions and gestures account for up to 60–65 percent of your communication.[174,175]

169 Alex Pentland. 2010. Honest Signals: How They Shape our World. MIT Press. Cambridge, Massachusetts.

170 Alex Pentland. 2010. To signal is human. American Scientist. 98:204–211.

171 Eric Jaffe. 2007. Mirror neurons: How we reflect on behavior. May. Psychological Science. https://www.psychologicalscience.org/observer/mirror-neurons-how-we-reflect-on-behavior.

172 Andrea Garcia Cerdan. 2017. Mirror neurons: The most powerful tool. Cognifit. 8 June. https://blog.cognifit.com/mirror-neurons/.

173 Sooyeon Kim. 2013. Neuro-cognition and social-cognition: Application to exercise rehabilitation. Journal of Exercise Rehabilitation 9(6):496–499.

174 Kendra Cherry. 2019. Understanding body language and facial expressions. September 4. https://www.verywellmind.com/understand-body-language-and-facial-expressions-4147228.

175 Gretchen Foley and Julie Gentile. Nonverbal communication in psychotherapy. Psychiatry (Edgemont). 2010; 7(6):38–44.

Here is a summary of emotions and associated voice intonation, pace, and amplitude, and gesture signals.[176,177,178,179,180,181,182] Notice that emotions can be indicated by more than one signal, and signals associated with more than one emotion. Your ability to convey your intent, and assess the intent of those you communicate with, improves markedly if you are attentive to multiple signals.

- *Anger and Hostility.* Grimace and furrowed brow, clasped hands behind back, and hiding the trunk of the body by hunching forward and crossing arms and legs.
- *Approval, Friendliness, Happiness, and Optimism.* Smile, slightly turned-up mouth, and trunk of the body open and exposed.
- *Boredom, Impatience, Distraction, Concealing, and Frustration.* Hands clasped behind back, fidgeting, tapping fingers or toes, hunched forward, and looking away.
- *Confidence.* Emphatic tones, fluid speaking, circulating among a crowd, listening intently, and driving conversations with questions.
- *Confusion.* Eyebrows drawn together, head tilting to the side, and a distant gaze.
- *Control and Aggressiveness.* Hands on hips.
- *Defensive.* Crossed arms and legs.

176 Kendra Cherry. 2019. Understanding body language and facial expressions. September 4. https://www.verywellmind.com/understand-body-language-and-facial-expressions-4147228.

177 Alexander Todorov, Sean Baron, and Nikolaas Oosterhof. 2008. Evaluating face trustworthiness: a model based approach. Society for Cognitive Affective Neuroscience 3(2):119-27.

178 Paul Ekman. 2009. Darwin's contributions to our understanding of emotional expressions. Philosophical Transactions of the Royal Society of London, Bulletin of Biological Science 364(1535):3449–51.

179 Thomas D'agostino and Carma Bylund. 2014. Nonverbal accommodation in health care communication. Health Communication 29(6):563–73F.

180 Frank Marchak. 2013. Detecting false intent using eye blink measures. Frontiers in Psychology. 4:736. 11 October. https://www.frontiersin.org/articles/10.3389/fpsyg.2013.00736/full.

181 Gretchen Foley and Julie Gentile. Nonverbal communication in psychotherapy. Psychiatry (Edgmont). 2010;7(6):38–44.

182 Tanya Vacharkulksemsuk, Emily Reit, Poruz Khambatta, Paul Eastwick, Eli Finkel, and Dana Carney. 2016. Dominant, open nonverbal displays are attractive at zero acquaintance. Proceedings National Academy of Sciences, USA 113(15):4009–14.

- *Disapproval and Distrust.* Frown, tightly pressed lips, and slightly turned down mouth.
- *Discomfort and Anxiety.* Looking away from the listener, frequent blinking, biting lips, and hiding the trunk of the body by hunching forward and crossing arms and legs.
- *Excitement and Surprise.* Eyes wide open, raised eyebrows, and smile.
- *Interest.* Speaking more than listening, listening intently, fluctuation in voice amplitude and pitch, looking directly into the eyes, and sitting or standing up straight.
- *Openness.* Trunk of the body open and exposed.
- *Sadness.* Frown and slightly turned down mouth.
- *Trustworthiness.* Slight raise of the eyebrows and a smile.

Even before COVID-19, social distance was another component of face-to-face communication.[183] Choosing the wrong distance can trigger feelings of anxiety, aggressiveness, confusion, defensiveness, distraction, and hostility.

Typically, people in an intimate relationship are comfortable with distances of 6 to 18 inches. Family and close friends tolerate 1.5 to 4 feet. Acquaintances and colleagues accept 4 to 12 feet, and comfortable public interactions (e.g., public speaking, presentations) are 12 to 25 feet.

Social distance is modified by culture. For example, Latin Americans are typically more comfortable with lesser distances than North Americans. Observe the distances maintained by other cultures to avoid unintentionally triggering negative emotions.

Communicate with Visuals

Visual tools (i.e., graphs, charts, maps, and diagrams) can improve your communication. The human brain is sensitive to any information that is a symbolic representation of a series of events.[184] You process visuals 60,000

183 Edward Hall. 1963. A system for the notation of proxemic behavior. American Anthropologist. October. 65(5):1003–1026.

184 Alan Paivio. 2006. Dual coding theory and education. Draft chapter for the conference on Pathways to Literacy Achievement for High Poverty Children. University of Michigan School of Education. 29 October.

times faster than words.[185] And you remember visuals better than words, even better than visuals and words.[186]

Within seconds, someone viewing the following figure can come to two conclusions: 1) the Earth's temperature is rising, and 2) the rise in temperature is not due to increased solar irradiance.

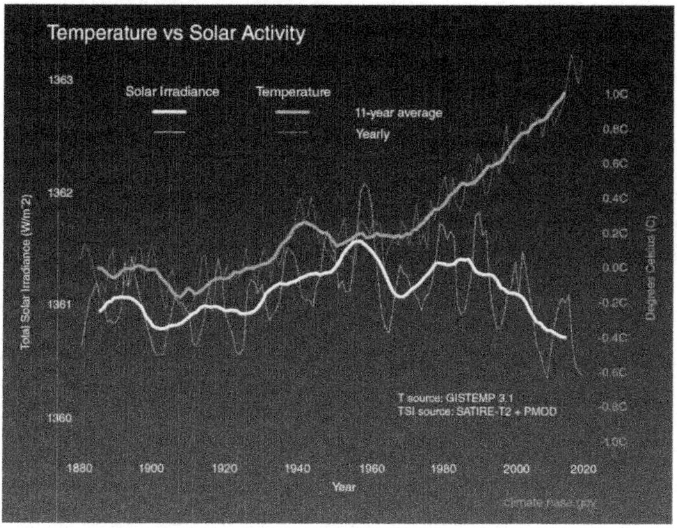

Source: NASA 2021[187]

You can thank your brain for this ability to quickly interpret visuals. The visual cortex is the primary cortical region of the brain that receives, integrates, and processes visual information relayed from the retinas. It lies in the occipital lobe of the primary cerebral cortex, which is in the most posterior region of the brain in both the left and right hemispheres.

The right hemisphere of the brain also powers your visual interpretation. The hemisphere is dedicated to visual processing, seventy percent of your

185 Krista Neher. 2013. Visual Social Media Marketing – Harnessing Images, Instagram, Infographics and Pinterest to Grow Your Business Online. https://www.slideshare.net/kristaneher/neher-biz-tech/5-Your_brain_processesimages_60000_timesquicker.

186 Joan Snodgrass, Rochelle Volvovitz, and Eleanor Walfish. 1972. Recognition memory for words, pictures, and words and pictures. Psychonomic Science 27(6):345–347.

187 National Aeronautics and Space Administration. 2021. https://climate.nasa.gov/climate_resources/189/graphic-temperature-vs-solar-activity/

sensory receptors are in your eyes, and ninety percent of information transmitted to your brain is visual. You identify images seen for as little as 13 milliseconds.[188]

Visuals can convey more information than words and facial expressions. Your brain quickly absorbs large, complex visual data. Visuals can affect emotions and leave impressions faster and more accurately than words.[189] Visuals are universal, apparent to anyone who has tried to navigate the streets of foreign countries.

Which communicates better and more quickly? This statement:

Atmospheric carbon dioxide levels have cycled over the last 800,000 years, but they have never exceeded 300 parts million until 1950. The current level is now more than 400 parts per million.

Or the figure below? Science tells us the figure is a better communicator than the statement.

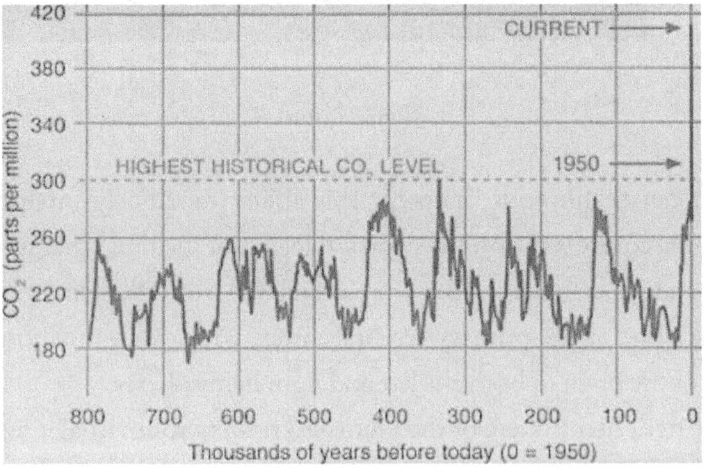

Source: NASA 2021[190]

188 Anne Trafton. 2014. In the blink of an eye. https://news.mit.edu/2014/in-the-blink-of-an-eye-0116.

189 Dawn McBride and Barbara Dosher. 2002. A comparison of conscious and automatic memory processes for picture and word stimuli: a process dissociation analysis. Consciousness and Cognition 11(3): 423–460.

190 National Aeronautics and Space Administration. 2021. https://climate.nasa.gov/vital-signs/carbon-dioxide/

Your brain is trained to decipher all elements of a visual image at once, connecting your memories and emotions. This facilitates getting attention, conveying a message, and eliciting an emotional response.

Visuals are perceptually more distinct from one another than are words, increasing their chance for retrieval. Visuals are also thought to assess meaning more directly than words.[191]

You should consider visuals when sharing large and complex information. Especially information you hope will be remembered in support of healthy lands and waters.

Persuasion

At some time or another you will want to persuade someone of your viewpoint. To change their beliefs. Knowing your audience is a good start (see above). Now you need to understand how to sell. Not sell in a pejorative sense but sell in that you have something valuable to offer others.

Motivational Interviewing

Adam Grant, author and professor at the esteemed Wharton School of Business at the University of Pennsylvania, no longer believes it is his place to change anyone's mind. He came to this perspective thoughtfully, through experiencing people shutdown or become more adamant when arguing the opposite of their beliefs. Instead, he tries to understand their thinking and openness to rethinking. Only then does he apply persuasion in the form of motivational interviewing[192,193].

I do not counsel restricting your persuasion only to those ready to rethink. The deaths of tens of millions of people each year due to environmental

191 William Hockley and Tyler Bancroft. 2011. Extensions of the picture superiority effect in associative recognition. Canadian Journal of Experimental Psychology 65(4):236–244.

192 Adam Grant. 2021. The science of reasoning with unreasonable people. https://www.nytimes.com/2021/01/31/opinion/change-someones-mind.html?ck_subscriber_id=864066662&utm_source=convertkit&utm_medium=email&utm_campaign=The+Core+Pillars+of+Immune+Health%20-%205357582

193 William Miller and Stephen Rollnick. 2013. Applications of Motivational Interviewing. 3rd edition. Guilford Press.

causes, impacts increasingly exacerbated by climate change and biodiversity loss, is too compelling. But Adam Grant's point is well- taken and motivational interviewing is an important technique.

Motivational interviewing, a technique previously applied to substance abuse problems, can help others find their intrinsic motivation to change. The technique uses open-ended questions, careful listening, and reflecting answers to help others more clearly see their thoughts. As typically practiced, only if they express a desire to change do you provide guidance.

Psychologists find that people's thinking becomes less extreme and more open when someone listens carefully and calls attention to the nuances of their beliefs. Social scientists find asking people how their belief works in practice is more effective than asking why they hold that belief. A skilled motivational interviewer listens for change talk, a desire, ability, or commitment to shifting a belief, and asks people to elaborate.

6 Principles of Persuasion

How to get people to say yes has been studied for more than 60 years. Robert Cialdini is a leader in the field and through years of research he identified 6 principles of persuasion.[194]

1. *Reciprocity.* People feel obligated to give back to those who have given them something. Give something personal and unexpected before someone gives to you.
2. *Scarcity.* People want more of things in limited supply. Show what is unique about your proposition and what they lose if they do not accept.
3. *Authority.* People follow credible experts.
4. *Consistency.* People like their new actions to be consistent with their previous actions.
5. *Liking.* People preferentially say yes to those who are like them, who pay them compliments, and who cooperate toward mutual goals.

194 Robert Cialdini. 1993. Influence: The Psychology of Persuasion. William Morrow and Company. New York, New York.

6. *Consensus.* People are more likely to say yes when consistent with the actions of others. The influence of others is stronger when the person is uncertain.

Apply the 6 principles of persuasion, ascertaining which work best for you and those you want to influence.

Lessons From Copywriting and Fundraising

Copywriting and fundraising are applications of motivational interviewing and Cialdini's principles of persuasion. Both disciplines must persuade others to do something - buy in the case of copywriters and give in the case of fundraisers.

Copywriting

Copywriters are salespeople and persuaders. They are the drivers behind the majority of product and service sales worldwide. Good copywriters help sell a lot of products and in the process make a good living. The best copywriters have perfected the craft of persuasion.

To persuade others of what you are selling, you must do three things – get attention, communicate, and persuade someone to act. [195]

Begin with getting attention. You have 3 to 5 seconds to gain the audience's attention, or you lose them. Getting attention is the job of a headline. Your headline, written or verbal, should appeal to the audience's self-interest, giving them news or offering useful information. Your headline should include the strongest benefit and powerful attention-getting words and phrases like how to, why, quick, easy, last chance, guarantee, results, proven, and save.

An effective headline selects your audience, delivers a complete message, and leads your audience to the bulk of the information. Know your audience, and in your headline speak directly and only to them. Make your headline a

195 Robert Bly. 2005. The Copywriter's Handbook: A Step-by-Step Guide to Writing Copy That Sells. Holt Paperbacks. New York, New York.

complete statement. Four out of five people will read the headline and skip the remaining information.

To appeal to most of your audience your headline should minimally include why they should care (e.g., the challenge), the solution, and how they can help. Arouse their curiosity to convince the audience to keep reading or listening. You can use humor, intrigue, and mystery, ask a question, or make a provocative statement. Or you can promise a reward, news, or useful information.

The next step in copywriting persuasion is to communicate. Speak in language that will be understood and use a conversational style. Offer something of interest. Organize your information. Restrict yourself to one major idea and get to its point in your lead paragraph. Write or speak in short sections and use short, crisp sentences that vary in length. Avoid unnecessary words and provide details. Your lead paragraph should get to the point of your big idea.

Finally, to persuasion itself. To persuade others of the value of product or service, you must focus on the benefits to your audience rather than its features. Show the audience they have a need or problem. Illustrate how your solution fulfills their need or solves their problem.

Provide evidence proving your solution will be effective. Tell your audience why your solution differs from other solutions and why you are the best at implementing it. Close by asking your audience to learn more or make a purchase.

Fundraising

Another area to look to for persuasion guidance is fundraising. Fundraising is not too different than copywriting except the end goal is a purpose rather than a product or service. Key elements of fundraising are clarity of purpose, constructing an urgent and inspirational message, and building relationships.

Your organization's clarity of purpose comes from your vision and mission. Before launching a fundraising campaign, review the vision and mission with your staff to ensure everyone knows what the organization is trying to do, and can capably discuss it with supporters. Talk through the details with your fundraising, communication, and marketing personnel. Everyone should understand their roles and responsibilities.

Anchored by your purpose, be clear about what you and your organization have to offer. Is it the chance for underprivileged children to learn and grow? Clean drinking water in less developed countries? Protected lands and waters that provide recreation, flood protection, and wildlife habitat?

Develop a value proposition, a statement that answers the *why* someone should support you and your organization while convincing them how your approach is more effective than alternatives. Prepare a story illustrating your value proposition.

Equipped with your value proposition, meet with an existing or potential supporter, and use active listening to determine what is important to them. Ask questions and seek advice while trying to build common purpose.

When you find that common purpose, inspire your potential supporter by tapping into their deeper need for fulfillment. What makes them feel like they are making a difference? Saving a child's life? Saving a beautiful place or protecting an endangered species? Be clear about the impact of their support.

Lastly, but no less importantly, ensure that your supporters share in those impacts. You can create a role for them in your efforts, invite them to celebrate success with you, or officially recognize their support.

Regardless of how you persuade, keep in mind that people generally respond first with their emotions and later rationalize their decision. Be authentic and passionate in your delivery. Start with a complete, attention-getting opening, communicate effectively, and persuade in a way that logically fulfills your audience's need.

DE-ESCALATE STRESSFUL SITUATIONS

It's not stress that kills us; it is our reaction to it.

Hans Selye, pioneering endocrinologist known for his work on the hypothetical, non-specific response of an organism to stressors

Twenty years of worldwide research involving more than 100,000 people, 2,200 projects, and hundreds of organizations revealed that a key skill of effective leaders is the capacity to address emotionally and politically risky issues.[196] This skill predicts with nearly 90 percent accuracy which projects will fail months or years in advance. Specifically, success or failure depends on whether people could hold five critical conversations encompassing scope and schedule, team member performance, and failed leadership.

Conflict and Stress

Conflict and attendant stress are increasingly common.[197] More situations seem to be complex and consequential. Will I lose my job if I make a mistake? Why does that person always argue with me? Why won't my supervisor talk to me? My workload is too large! I do not have any control over my work life! If I fail, life on Earth will end!

196 Kerry Patterson, Joseph Grenny, Ron McMillan, and Al Switzler. 2011. Crucial Conversations: Tools for Talking When Stakes are High. McGraw-Hill.

197 Mental Health Foundation. 2018. Mental health statistics: Stress. 4 June. https://www.mentalhealth.org.uk/statistics/mental-health-statistics-stress.

Conflict, and anticipation of other future conflicts, triggers stress. Your body temperature rises, you sweat, your pulse quickens, breaths shorten, muscles tense, and pupils dilate. You are experiencing a fight or flight response.[198] Your autonomic nervous system is regulating your bodily functions and acts for the most part unconsciously.

Your sensory organs send a signal to your amygdala for processing, which alerts the hypothalamus. Your hypothalamus via the autonomic nervous system causes the adrenal glands to release epinephrine (aka adrenaline). Epinephrine causes your heart to beat faster in order to deliver blood to the heart, vital organs, and muscles.

Small airways in the lungs open wide so you can take in more oxygen for the brain. Your senses sharpen. Blood sugar and fats are released from temporary storage sites into the bloodstream providing energy to all parts of the body. Faced with prolonged threat, your hypothalamus prompts your adrenal glands to release cortisol, keeping you alert.

A fight or flight response is an elegant and critical reaction to danger, but a dangerous reaction to a difficult but nonthreatening situation. [199, 200] Chronic stress does not allow your body to recover from the effects of epinephrine and cortisol. Your immune, digestive, cardiovascular, sleep, and reproductive systems are disrupted. Blood pressure remains high, artery-clogging deposits form, and brain changes contribute to anxiety, depression, and addiction. Your susceptibility to kidney disease, type II diabetes, obesity, and Parkinson's disease increases.

Stress can also affect your decision-making.[201] Stressed individuals are less able to override their habitual decision making and adapt to changing

198 Harvard Health Publishing, Harvard Medical School. 2011 (update 2018). Understanding the stress response. 1 May. https://www.health.harvard.edu/staying-healthy/understanding-the-stress-response.

199 Nicoletta Lanese. 2019. Fight or Flight: The sympathetic nervous system. 9 May. Live Science. https://www.livescience.com/65446-sympathetic-nervous-system.html.

200 National Institute of Mental Health. (no date). 5 things you should know about stress. NIH Publication No. 19-MH-8109.

201 Barbara Isanski. 2010. Under pressure: Stress and decision making. 1 August. Psychological Science. https://www.psychologicalscience.org/observer/under-pressure-stress-and-decision-making.

circumstances.[202,203] And chronic stress can make you focus more on potential rewards than on potential risks.[204]

Stress Assessment

Take this stress self-assessment modified from Patterson et al.[205] Do you recognize any of your behaviors.

1. Do you avoid situations that might bring you into contact with people with whom you are having a problem?
2. Do you postpone returning phone calls or e-mails because you don't want to deal with the person who sent them?
3. Do you try to change the subject when people bring up a touchy or awkward issue?
4. Do you hold back rather than give your full and candid opinion when dealing with awkward or stressful subjects?
5. Do you rely on jokes, sarcasm, or snide remarks to let people know you are frustrated?
6. Do you offer weak or insincere compliments to soften the blow when you have something difficult to bring up?
7. Do you exaggerate your side of the argument to get your point across?
8. Do you interrupt people or return the subject to where you want it when you seem to be losing control of a conversation?
9. Do you let others know if you think something is stupid?
10. When surprised do you say things like *Give me a break*! or *That's ridiculous*!?
11. Do you get angry and stop discussing others' points and make it personal?

202 José Soares et al. 2012. Stress-induced changes in human decision-making are reversible. Translational Psychiatry 2(7): e131.

203 Anthony Porcelli and Mauricio Delgado. 2017. Stress and decision making: Effects on valuation, learning, and risk-taking. Current Opinions in Behavioral Science. April 14:33–39.

204 Alexander Friedman et al. 2017. Chronic Stress Alters Striosome-Circuit Dynamics, Leading to Aberrant Decision-Making, Cell 171. 5: 1191–1205.

205 Kerry Patterson, Joseph Grenny, Ron McMillan, and Al Switzler. 2011. Crucial Conversations: Tools for Talking When Stakes are High. McGraw-Hill.

12. Do you get angry and make hurtful and insulting comments?

According to Patterson et al., if you answered yes to the first six questions you tend toward silence under stress. Answering yes to the first two questions indicates you are withdrawing from the situation. Answering yes to questions 3 and 4 indicates you are avoiding sensitive situations. Answering yes to questions 5 and 6 indicates you are masking, understating or selectively showing how you really feel.

If you answered yes to the second six questions you tend toward violence in stressful situations. Answering yes to questions 7 and 8 indicates you attempt to control others. Answering yes to questions 9 and 10 indicates you label people or ideas so they can be stereotyped and dismissed. Answering yes to questions 11 and 12 indicates you attack. You are not interested in the best idea, and use threats to ensure your idea is chosen.

De-escalating Stressful Situations

Silence and violence are not the answer. De-escalating stress is essential if leaders are to stay healthy, make good decisions, and work with others. De-escalating stress is essential if you want satisfied, creative, and productive teams. There is hope.

A recent study by Lees and Cikara found that people are more willing to compromise than we think.[206] The way forward includes avoiding stereotypes, recognizing others as individuals rather than *them*, and coming to the situation with a cooperative mindset.

How do you proceed knowing that stress is bad and stressful situations can be managed? Two practices can help. The first – mindfulness - helps you develop a cooperative mindset in preparation for a stressful situation. The second - managing afflictive emotions - helps you navigate the situation.

206 Jeffrey Lees and Mina Cikara. 2019. Inaccurate group meta-perceptions drive negative out-group attributions in competitive contexts. Nature Human Behavior. 2019. November.

Mindfulness

Mindfulness is not exotic or denominational but a quality you already possess. Mindfulness is a basic human ability to be fully present, aware of where you are and what you are doing.[207,208] It allows you to avoid overreacting and being overwhelmed by your physical, social, and emotional environment. Mindfulness is the time you suspend judgment and unleash your natural curiosity.

Many studies demonstrate the benefits of mindfulness. Mindfulness reduces your stress, anxiety, and depression and improves your focus and ability to hold information for processing.[209,210,211,212] Mindfulness increases your compassion, your willingness to accept different narratives, and your ability to regulate your emotions.[213]

Reducing stress, anxiety, and depression, improving focus, increasing compassion and receptivity to contrary thoughts, and regulating your emotions are all beneficial for successfully navigating stressful situations. How do you practice mindfulness? Through meditation.

There are two types of meditation that enable leadership in stressful situations. The first type of mediation - gathering the mind - is a prerequisite to the second type - contemplative meditation.[214]

Gathering the Mind Meditation

In preparation, create some time and space for regular, consistent meditation. Ten to 20 minutes twice a day is enough to obtain the benefits. Meditate

207 Janice Marturano. 2014. Finding the Space to Lead. Bloomsbury Press. New York, New York.

208 Mindful Staff. 2014. What is mindfulness? 8 October. https://www.mindful.org/what-is-mindfulness/.

209 Stefan Hofmann, Alice Sawyer, Ashley Witt, and Diana Oh. 2010. The effect of mindfulness-based therapy on anxiety and depression: A meta-analytic review. Journal of Consulting and Clinical Psychology 78(2):169–183.

210 Paul Grossman, Ludger Niemann, Stefan Schmidt, and Harald Walach, Harald. 2004. Mindfulness-based stress reduction and health benefits. Journal of Psychosomatic Research. July 57(1):35–43.

211 John Cloud. 2010. Time. Friday 6 August. http://content.time.com/time/health/article/0,8599,2008914,00.html.

212 Amishi Jha, Jason Krompinger, and Michael Baime. (2007). Mindfulness training modifies subsystems of attention. Cognitive, Affective, & Behavioral Neuroscience 7(2):109–119.

213 Daniel Goleman and Richard Davidson. 2017. Altered Traits: Science Reveals How Meditation Changes Your Mind, Brain, and Body. Avery. New York, New York.

214 Sakyong Mipham. 2003. Turning the Mind Into an Ally. Riverhead Books. New York, New York.

in a place that is comfortable, quiet, and clean and wear comfortable clothing. Check your mindset before beginning. If you are tense or anxious, take a few slow, deep breaths and concentrate on relaxing your body. Let your day disappear.

To begin, sit with the spine upright with a natural curve. If you sit on a cushion loosely cross your ankles. If on a chair keep both feet firmly on the floor. Rest your hands on your thighs or joined loosely in your lap with your arms and shoulders relaxed. Your chin should be slightly tucked, eyelids half shut, and your face and jaw natural and relaxed.

Place your mind on your breath and breathe normally. Tell yourself you will now let go of concepts, thoughts, and emotions. Focus on your inhalations and exhalations. When you lose your focus on your breath, and you will, take note and return. Do not grasp at concepts, thoughts, and emotions as they arise. Simply acknowledge them, observe, and return to your breath.

The more you practice, the easier it will be to hold your mind on your breath. Do not chastise yourself for being imperfect. Accept that meditation is helping you be mindful, and that mindfulness will reduce your stress, improve your focus, help you regulate your emotions, and increase your compassion and receptivity to contrary thoughts.

Contemplative Meditation

After a few weeks of gathering the mind meditation, you will have developed enough mindfulness to engage in contemplative meditation. The object of contemplative meditation is not the breath but concepts, thoughts, and emotions. Your goal is to examine a concept, thought, or emotion and ascertain its meaning. In the context of stressful situations, you want to develop the wisdom to de-escalate stress, cooperate, and find a compromise.

Begin as you would for gathering the mind meditation. After a few moments, shift your focus from your breath to a certain concept, thought, or emotion. For example, why am I so upset when John Smith disagrees with me about our priority, or why do I think Jane Smith's idea is wrong? Gather your mind and place it on those words. When you recognize you are thinking about something else, return to the object of your meditation.

Bring in other thoughts, images, and memories if they enrich your contemplation. As you continue, the words should fall away and their meaning or the experience becomes clear. At the end of your meditation let the meaning of your session become your new perspective. Bring that perspective to the stressful situation.

Gathering the mind meditation serves as your default. As little as eight weeks of consistent, regular meditation should reduce your stress, improve your health and focus, regulate your emotions, and increase your compassion and willingness to accept a new perspective. You can practice contemplative meditation as needed to improve your capacity to manage stressful situations and find solutions.

Managing Afflictive Emotions

Many of your stressful situations are attributed to afflictive emotions - obscuring mental factors that prevent the mind from establishing reality.[215] Anxiety, fear, anger, desire, depression, greed, lust, hatred, attachment, ignorance, and pride are afflictive emotions.

These mental states obscure reality and give rise to misconceptions - modes of thought at odds with the facts.[216] If you don't work with these intense emotions they grow and you act on them.[217] Your mind weaves from one idea to another, one emotion to another.

Managing your afflictive emotions with mindfulness practices will help you remain in dialogue and take turns thoughtfully and respectfully speaking and actively listening. You will bring to the situation a cooperative mindset with the goals of mutual purpose, understanding, and a path forward. You will hold conversations that couple your brain with others ensuring meaningful communication.[218]

215 Daniel Goleman. 2004. Destructive Emotions: A Scientific Dialogue with the Dalai Lama. Bantam Books. New York, New York.

216 Dalai Lama. 2006. How To See Yourself As You Really Are. Atria Books. New York, New York.

217 Sakyong Mipham. 2003. Turning the Mind Into an Ally. Riverhead Books. New York, New York.

218 Uri Hasson. 2017. This is your brain on communication. https://ideas.ted.com/this-is-your-brain-on-communication/. 27 January.

Path to Action

Patterson et al. devised a simple and effective practice for managing afflictive emotions and remaining in dialogue.[219] Paraphrased, the practice has four elements that reflect thought processing.

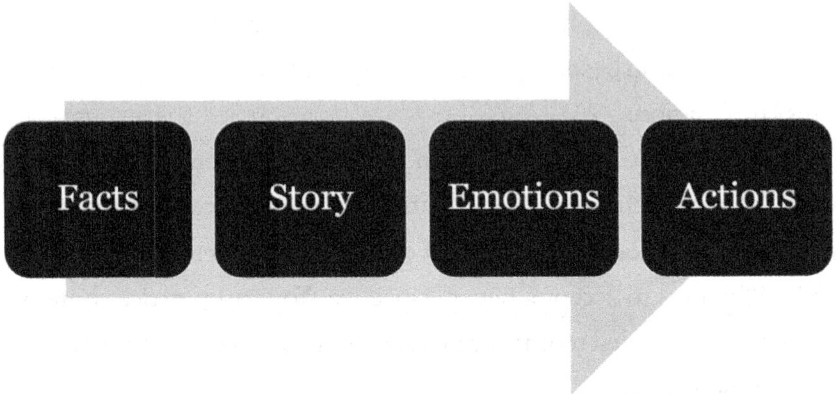

Source: Adapted from Kerry Patterson et al. 2011.

You can use the practice to structure your position so it is clear and understandable to the other person. But the more important use is to return to dialogue when a conversation has gone wrong.

Ask the other person to retrace their path. What facts form the basis for their position? How do they interpret those facts? How does their interpretation of the facts make them feel? Are their proposed actions consistent with their emotions? When probing the other person's facts be tentative and humble rather than forceful and assured.

Now state your path. Where do the two paths diverge? Begin by trying to reach agreement on the facts. Offer your own facts to demonstrate your openness to different perspectives. Beliefs are often construed as facts so be sure you know which you are presenting. Do not examine stories without coming to agreement on the facts.

219 Kerry Patterson, Joseph Grenny, Ron McMillan, and Al Switzler. 2011. Crucial Conversations: Tools for Talking When Stakes are High. McGraw-Hill.

Next, examine each other's stories. Do your stories accord with the facts? Are there other stories that accord with the facts? In many situations, there are multiple stories that accord with the facts. Your goal is to find a story that works for both of you.

Now examine each other's emotions. Do your emotions match the story? If your emotions differ examine the basis for those differences. Different belief systems and cultural contexts can produce differing emotions for the same story. Your goal is not to agree on a common emotion but to respectfully understand why you feel differently and how they affect your actions.

Finally, examine each other's proposed actions. Do the actions naturally follow from your emotions? From your stories? From your facts? Are other actions equally or more reasonable? Continue until you find a mutually acceptable solution or agree to disagree. If the latter, the person with authority makes the decision.

The path to action practice works equally well with groups like staff, board members, partners, rivals, and opponents. Explain the path to participants at the beginning of a meeting and always keep it visible. Visibility keeps the path in the forefront of everyone's mind so they can monitor themselves and others. Ask everyone to be cognizant of whether what they and others are saying is a fact, story, or emotion. And whether the proposed actions are in accordance with the facts, stories, and emotions.

The Benefits of De-escalating Stress

The benefits of being mindful, managing your afflictive emotions, and engaging in meaningful communication are significant.[220] Organizations with employees skilled in critical conversations:

- Respond 5 times faster to financial downturns.

220 Kerry Patterson, Joseph Grenny, Ron McMillan, and Al Switzler. 2011. Crucial Conversations: Tools for Talking When Stakes are High. McGraw-Hill.

- Are 67 percent more likely to avoid injury or death due to unsafe conditions.
- Save over $1,500 and an eight-hour workday for every critical conversation an employee holds rather than avoids.
- Substantially increase trust and reduce transaction costs in virtual work teams.
- Influence positive change in colleagues who are bullying, conniving, dishonest, or incompetent.

MANAGE GROUP DYNAMICS

Alone we can do so little; together we can do so much.

Helen Keller, blind and deaf American author,
political activist, and lecturer

What Are Group Dynamics?

Group dynamics refers to the interplay of people's social learning and behaviors operating within a group or between groups. The term was coined by social psychologist, Kurt Lewin, who established the Group Dynamics Research Center in 1945 at the Massachusetts Institute of Technology. His work revealed three fundamental findings for group dynamics and learning.[221]

First, groups have three types of leaders reflecting an increasing degree of control.

1. *Laissez-faire leaders* do not interfere, letting group actions take their own course. Groups with a laissez-faire leader are less productive than other groups.
2. *Democratic leaders* believe that all group members have an equal right and opportunity to contribute. Groups with a democratic leader are more cooperative and enjoy their experience.
3. *Autocratic leaders* are domineering, exert absolute power, and have little regard for the opinions or contributions of others. Group members

221 Kurt Lewin and Gertrud Weiss Lewin. 1948. Resolving Social Conflicts: Selected Papers on Group Dynamics. Harper and Brothers. New York, New York.

with autocratic group leaders tend to be more dissatisfied, aggressive, and apathetic.

Lewin's second fundamental finding was that group members are far more likely to change their habits if encouraged to discuss issues themselves and make their own decisions. Just being provided appropriate information, directions, and advice had little effect on changing habits.

Thirdly, most group members want to stay within the behavioral norms of the group. Only driving forces that are strong enough to break habits and challenge the interests and customs of the group will overcome resistance to change.

Groups must look at all the options for moving from the existing present to a desired future state, evaluate the possibilities of each, and decide on the best one to achieve change. Aiming for the desired goal and taking the straightest and easiest route is ineffective and counterproductive.

Build an Effective Group

Alex Pentland's group at the Massachusetts Institute of Technology Media Lab demonstrated that groups are smarter than individuals.[222] But working in groups is much more than sharing data or techniques. Collaborations have costs, especially when viewed in the short term. Groups require an investment of time, resources, and energy in building and maintaining relationships. Also, no one individual has control over outcomes and coordination and communication are challenging.

The best way to build an effective group is not to choose members because they are smart or accomplished, but because they are good communicators.[223] In effective groups:

222 Anita Woolley, Christopher Chabris, Alexander Pentland, Nada Hashmi, and Thomas Malone. 2010. Evidence for a Collective Intelligence Factor in the Performance of Human Groups. Science. Vol. 330, 29 October 2010, pp 686–688.

223 Pentland, Alex. 2012. The New Science of Building Great Teams. Harvard Business Review, http://hbr.org/2012/04/the-new-science-of-building-great-teams/ar/pr.

- Everyone on the team talks and listens in equal measure.
- Contributions to the group are short.
- Members face one another.
- Conversations and gestures are energetic.
- Members connect directly with one another not just with the leader.
- Members engage in informal conversations outside formal meetings.
- Members periodically explore outside the group and bring information back.

Socializing outside the meeting is critically important to group performance, accounting for more than 50 percent of positive changes in communication patterns. Socializing allows group members to exchange experiences and test ideas in a less formal setting.

Group meetings should be frequent (e.g., weekly), but brief and preferably face-to-face. More than a third of the variation in a group's performance is accounted for by the number of face-to-face exchanges. The ideal number of exchanges is as many as dozens per hour. More than that tends to decrease performance.

Video conferences and telephone calls are less effective than face-to-face meetings and increasingly ineffective as the number of participants increases. Email and texting are the least effective forms of communication.

Group Dynamics

Group dynamics play a significant role within and between individuals and organizations. Groups are comprised of people with different ideas, motivations, backgrounds, and different agendas. Most groups have a leader, and if not, look to appoint a leader. As a leader, you help develop a cultural bond that maintains group cohesiveness.

The more difficult it is to obtain group membership the more cohesive the group. Groups also tend to become cohesive when they are in intense competition with other groups or face a serious external threat to survival. Smaller groups and those who spend considerable time together also tend to be more cohesive.

Leading a group requires empathy - the ability to sense others' feelings and perspectives. Empathetic leaders take an active interest in group member concerns and pick up cues to what is being felt and thought.[224] Empathy is positively related to performance.[225]

Leading a group also requires organizational awareness - the ability to recognize the power relationships, emotional currents, networks, influencers, and dynamics in the organization or among organizations.[226]

Group Development

The basis for diagnosing group dynamics is rooted in group development. Research by educational psychologist Bruce Tuckman identified a five stage development process leading to high performing groups.[227]

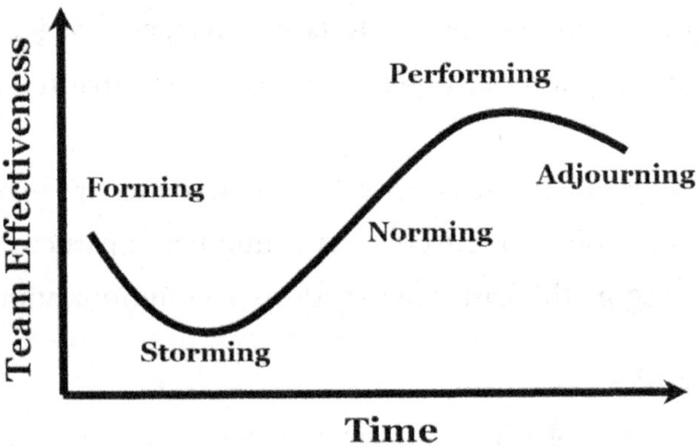

Source: ER Services depiction of Tuckman 1965.

224 Richard Boyatzis. 2016. How to Develop Empathy When it Doesn't Come Naturally. https://www.keystepmedia.com/develop-empathy-doesnt-come-naturally/#.WCxyVpI6R8c. 15 November.

225 William Gentry, Todd Weber, and Golnaz Sadri. Empathy in the workplace: A tool for effective leadership. Center for Creative Leadership.

226 Daniel Goleman. 2016. Can you read the room? Social awareness at work. https://www.linkedin.com/pulse/can-you-read-room-social-awareness-work-daniel-goleman/. 21 November.

227 Bruce Tuckman. 1965. Development sequence in small groups. Psychological Bulleting 63(6):384-399.

In the *forming stage*, groups orient themselves. Members test the boundaries of interpersonal and task behaviors. They develop dependency relationships with leaders, other group members, and pre-existing standards. Group members want to understand their benefits, what is expected of them, and their fit.

The second stage – *storming* - is characterized by interpersonal conflict and polarization. Members are inclined to resist group influence. Group performance typically decreases especially if members are not acquainted with each other. Group leaders must encourage members to resolve conflicts, be receptive to new ideas, and focus on the goal.

In the third stage – *norming* - group cohesiveness improves, new standards evolve, and roles are clarified and accepted. Intimate, personal opinions are expressed and accepted. Group members cooperate and focus on the goal.

In the fourth stage – *performing* - the group's interpersonal structure supports activities. Roles become flexible and functional and group energy is directed to pursuit of the goal. The group is mature, organized, and performing at a high level. Any problems are addressed constructively.

In the last stage – *adjourning* - the group's goal has been accomplished. Process and results are documented, the group celebrates its accomplishments, and members are assigned to other tasks.

Develop a Strategy to Evaluate Group Productivity

In 1970, Roger Harrison, an expert in team development and the design of self-managed learning systems argued that the depth of emotional involvement in an organizational change process determines the intervention strategy.[228] Harrison recommended applying deep diagnosis, reflection, and appreciation in advance of context-based organizational intervention.[229] Implementation would then occur at a level no deeper than needed to solve the problem and no deeper than the organization's commitment to resolution.

228 Roger Harrison. 1970. Choosing the depth of organizational intervention. *Journal of Applied Behavioral Science* 6(2):181–202.

229 Roger Harrison. 1995. Choosing the depth of organizational intervention, pp. 13–32, In R. Harrison (Ed.) *The Collected Papers of Roger Harrison*. McGraw-Hill. London.

In 1994, Roger Schwartz, organizational psychologist, author, and skilled facilitator, applied Harrison's theory to groups.[230] He used six levels of intervention.

1. Goals, values, and practices.
2. Roles and responsibilities.
3. Work processes.
4. Leaders' challenges and issues.
5. Interpersonal.
6. Intrapersonal.

In 2001, the Leadership Institute of Seattle shared the waterline model, an adaption of work by Harrison, Scheer, and Short. The model had four levels of intervention.[231,232,233]

1. *Structure* - purpose and goals, decision-making, and roles and responsibilities.
2. *Group* - equality, inclusiveness, behavior, outlook, and consistency.
3. *Interpersonal* - acceptance, respect, understanding, receptivity, and harmony.
4. *Intrapersonal* - attention, participation, and contribution.

Since then, the waterline model has been applied to diagnosing organizational change, team building, group dynamics, leadership training, and therapy.[234]

230 Roger Schwartz. 1994. *The Skilled Facilitator: Practical Wisdom for Developing Effective Groups*. Jossey-Bass. San Francisco.

231 Leadership Institute of Seattle. 2001. Waterline model of diagnostic and therapeutic intervention. (Adapted from Roger Harrison, J. Scherer, R.R. Short). 2001. Kenmore, WA: unpublished.

232 Abbey Louie. 2017. Dive In Answers Are Below the Waterline. https://leadx.org/articles/dive-answers-waterline/. 8 June.

233 Community Consulting Partnership. (accessed 20 December 2019). Diagnosing human systems: The waterline model. Seattle, Washington.

234 National Outdoor Leadership School. 2009. 2009 NOLS Leadership Educator Notebook: A Toolbox for Leadership Educators. (Eds. John Gookin and Shari Leach). 5th Edition. Lander, Wyoming.

Diagnose Group Dynamics

The waterline model helps you lead a cohesive, productive group by serving as a road map to averting problems. Conversely, the model can help you diagnose the cause of an incohesive, unproductive group. In accordance with Harrison's work you begin at the least emotional level and only proceed to deeper levels as needed.

As a leader, even if you are not the person in charge, you can affect the group's cohesiveness and productivity. You need to shine a light on the way forward, adopt effective communication methods, and acknowledge the responsibility of the person in charge to model behaviors that influence group behavior. Ask questions to draw out information hidden by reticent group members or groupthink. Help create a safe environment where group members can challenge each other.

If you are in charge, you will enhance the group's ability to be cohesive and productive if you begin the first meeting with some type of ice-breaker exercise designed to ease tension or relieve formality. An effective exercise is related to the group's goal, group work, varying roles, and the opportunity for everyone to be a leader. For example, the first group exercise could be a review of a draft work plan. Or construction of a group charter encompassing the structural topics below if the group is a known quantity.

Groups are complex and there are many ways cohesion and productivity can be impacted. The first meeting is key. Also critical is keeping structure, especially expectations, front of mind. A brief review of structure at the beginning of each meeting will help minimize group dysfunction.

Waterline Model

For diagnosing group dynamics, I recommend an adapted waterline model.

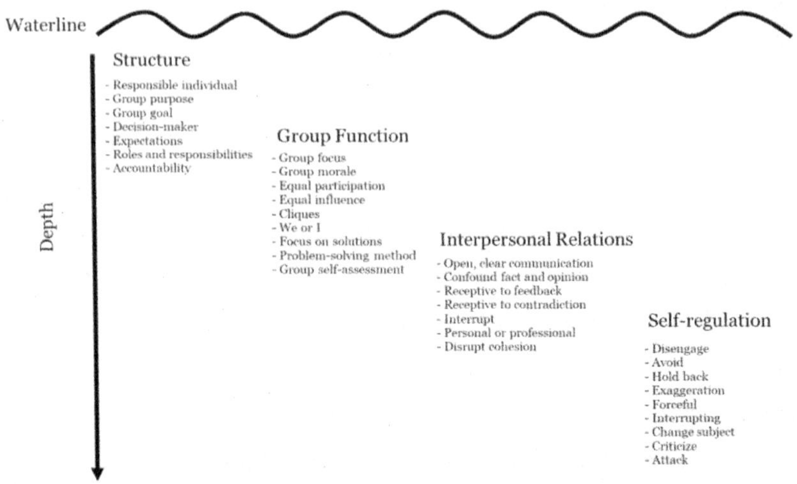

Structure

Structure establishes the group's clarity and alignment. Do not be shy about leading and enforcing group behavior. Most people want and need structure to participate in a group process. Reinforce that a cohesive group of good communicators is smarter and more productive than any individual.

Structure is your most fundamental tool for achieving your goal. Use the following questions and guidance to diagnose structure.

Who is responsible for the group's activities?

The group leader either owns the challenge/opportunity or was assigned to lead the group by a supervisor who does. Emphasize that you will be a democratic leader ensuring everyone has an opportunity to provide input and listening will be without bias.

Why was the group formed?

The group was formed to address an organizational or inter-organizational challenge or opportunity. Everyone was invited with the expectation that they had something valuable to offer. Do not form a group without a clear reason.

What is the group's goal?

The group's goal is to resolve the challenge or realize the opportunity. Be detailed enough so that the group fully understands the challenge or opportunity and knows what success looks like.

Who will make the decisions and how will they be made?

There are two alternatives to this question. If you own the problem/opportunity, then you will make the decisions. If you are the proxy for the owner, then you will present the group's findings with a preferred recommendation to the owner who will then make the decision.

Regardless of the circumstances inform the group that no decision will be made until after considering all the options for achieving the goal. Be suspicious of unanimous group decisions early in the process. They could be indicative of groupthink - making decisions as a group in a way that discourages creativity or individual responsibility. Ironically, highly cohesive groups are subject to groupthink.

What are the expectations of the group?

The expectation of the group and the individuals who comprise the group is to focus on the goal and contribute to group cohesion and productivity. Speak and listen in equal measure. Forbid personal criticism. Value each other's perspectives. Apply active listening skills and trust in the process.

What are the individual roles and responsibilities?

In some instances, group members will have specific roles. For example if the challenge is developing a natural resource management plan, then the group may consist of plant, wildlife, ecosystem, and conservation experts. Each will be expected to lead their domain. All group members should be encouraged to contribute their perspective regardless of the topic under discussion. Collective intelligence is enabled by diversity of thought.

How are people held accountable?

Individuals will be accountable to the group but held accountable by the leader. In group meetings the leader will manage the conversations ensuring they are focused on the goal and adhere to expectations. Extra-meeting work will be assigned through mutual agreement with clear outcomes, required resources, and adequate time.

At the structural level your goal as leader is to ensure group individuals understand who is responsible for the group's activity, why the group was formed, their goal, decision-making, individual and group expectations, and roles, responsibilities, and accountabilities. If each of these elements is understood and a problem remains, then you should advance diagnosing and addressing group function.

Group Function

This level focuses on group dynamics - the influential actions, processes, and changes that take place in groups.[235] Aspects include conflict, competition for social status, aggressive advocacy, reluctant compromise, and pressure on individuals.[236] Everyone in a group is influenced by group dynamics. Leaders should model expected behavior.

Use the following questions to diagnose group function.

Is everyone focused on the goal?

All group activities and individual contributions should be recognizably connected to pursuit of the goal.

How is group morale?

Group members should be happy, motivated, and committed to group activities and problem-solving. Unhappy and unmotivated individuals will be unproductive and impact group cohesion and productivity.

235 Donelson Forsyth. 2014. Group Dynamics (6th ed.). Wadsworth Cengage Learning. Belmont, California.

236 Jose Marques and Dario Paez. 1994. The black sheep effect: Social categorization, rejection of ingroup deviates and perception of group variability. European Review of Social Psychology 5(1):37–68.

Is everyone participating or do some dominate and others withdraw?

Group members should speak and listen in equal measure. Individuals prone to talking should help the group leader draw out reticent individuals.

Do individuals have equal and appropriate levels of influence?

Individuals bring their workplace status to a group. Less senior individuals tend to defer to more senior individuals of their organization and other organizations. Some of the inequity can be addressed by structuring the group to minimize supervisor/staff situations. Also, you can treat individuals in a way that signals that everyone's contributions are given equal weight.

Have cliques formed?

A clique is a small group of people with shared interests or beliefs. Cliques are inevitable and relatively harmless unless they monopolize the conversation or intimidate people from contributing.

Do individuals fail to speak for themselves and instead say we?

Saying *we* instead of *I* signals a timidity to express an individual opinion. *We* also suggests some undefined collective has reached a conclusion. Other members of the group are unable to assess the validity and the weight of the statement without more detail.

Does the group focus on problems or blame others rather than focus on opportunities?

Reviewing the story behind a challenge or an opportunity is instructive. Then the focus must be on solving the problem or realizing the opportunity. Stuck on problems or blaming someone not in the group disrupts cohesion and impedes productivity.

Does the group have a method for solving problems?

Expectations are the framework for problem-solving. Every situation, every group, is different. As the group forms, it will evolve a method for solving problems. For example, some groups prefer discussion whereas others prefer exercises or talking to stakeholders. Some groups prefer that every conversation involves the whole group whereas others prefer subgroups. Subgroup discussions should be brought back to the whole group.

Does the group regularly assess itself in a way that leads to improvement?

Groups are dynamic. Every individual in a group has a responsibility to demonstrate leadership. This includes a willingness to assess the group and themselves and to share their assessment with the whole group. Keep the assessments focused on improvement and not blame. Assessments that recognize a problem with group dynamics should be accompanied by recommendations for how to address the problem.

As a leader, you must regularly assess the groups' dynamics. Revisiting the goal and expectations is common during the early stages of a group, even if everyone said they understood.

Periodically ask the group what they think about the group's dynamics. Poor group dynamics should be addressed as soon as they arise.

Engage the group in appreciative inquiry. What is the group doing when it is working well? What are you and other individuals doing when the group is working well? Acknowledge conflict and explore it using the path to action described in the chapter on de-escalating stressful situations.

Disproportionate contributions to the group's pool of knowledge are at the heart of many group dynamics problems. Regularly reinforce that the group will have more ideas and make better decisions if everyone contributes their ideas and perspectives.

Do not allow people to hide behind *we.* Reinforce that all contributions have value and opposing perspectives are to be valued. Give specific tasks and roles to withdrawn or reticent individuals. Ask dominant individuals to mentor others by encouraging their contributions and demonstrating respect for their perspectives.

Interpersonal Relationships

If group problems persist the next level of intervention is interpersonal relationships. An interpersonal relationship is a strong, deep, or close association or acquaintance between two or more people. The relationship may be brief or enduring. Interpersonal relationships are important for employee satisfaction and better decision-making.

Fundamental interpersonal skills include being a good listener, communicating clearly, collaborating with others, working creatively with others, and not criticizing, condemning, and complaining.[237] The quality and quantity of interpersonal relationships affect individual health and well-being.[238] In a group, the interpersonal relationships of group members are interdependent.

This level encompasses attitudes, beliefs, and perceptions among two members of the group. The behavior of each member affects the other. The interpersonal relationship between two group members also affects the group.

Conflict may arise from a preexisting antagonistic relationship or from a new relationship among members with differing belief systems. Competition for social status and closed-mindedness can also create conflict.

You can use the following questions to diagnose interpersonal relationships.

Is an individual communicating openly and clearly with another individual?

A speaker has a responsibility to communicate honestly, ensure others understand, and accept feedback. A listener has a responsibility to listen without judgment, clarify understanding when needed, and provide feedback.

Is a pair of individuals confounding fact and opinion?

Facts deserve more weight than opinions in decision-making. Portraying an opinion as fact or perceiving a fact as an opinion creates a false sense of certainty and increases risk of failure. Opinions are critical to deliberations but should be clearly identified as such by speakers and listeners.

237 National Research Council. 2011. Assessing 21st Century Skills: Summary of a Workshop. J.A. Koenig, Rapporteur. Committee on the Assessment of 21st Century Skills. Board on Testing and Assessment, Division of Behavioral and Social Sciences and Education. Washington, DC. The National Academy Press.

238 James House, Karl Landis, and Debra Umberson. 2003. Social relationships of health, pp. 218–26, In Peter Salovey and Alexander Rothman (Eds.). Social Relationships and Health. Psychological Press. New York, New York.

Does an individual accept a contrary opinion and feedback from another individual?

For an individual to contribute to a group they must be willing to entertain a different perspective. For a group to function all individuals must be able to do so.

Is an individual regularly interrupting another individual?

Interrupting another individual is disrespectful, and interferes with adding information to the pool of knowledge.

Is the interaction between two individuals personal rather than focused on the group's goal?

Personal interaction is expressed as challenging someone's right or competence to speak.

Is the interaction between two individuals disrupting group cohesion?

Conflict between two individuals disrupts the group's focus. So too, contribution or temporal dominance by two individuals interferes with broader contributions whether the dominance is conflict or accord.

Return to the established expectations if faced with interpersonal dysfunction. Respect all perspectives, stay positive and focused on the goal, and trust in the decision-making process. Reinforce the need to work together to reach the best decision. Emphasize active listening and good communication skills. Do not call out any specific individuals in front of the group.

Act outside the group if reviewing expectations with the group fails. Collect information from the two individuals with the dysfunctional relationship. Speak to each independently tracing their path to action. Determine where the difference occurs.[239] Help them work through their difficulties by providing feedback to each. Emphasize the importance of the group and its

239 Steve Duck. 1990. Out of the Frying Pan and Into the 1990s. Journal of Social and Personal Relationships 7(1): 5–28.

goal. Bring the two individuals together and mediate a meeting of the minds and an agreement to act in accordance with the group's expectations.

Self-regulation

Self-regulation is a fundamental aspect of social and emotional intelligence.[240] The ability to control one's impulses, to think before you speak and react, to express yourself appropriately. Self-regulation is intertwined with other social and emotional skills such as self-awareness, self-motivation, and empathy. Daniel Goleman, a psychologist that described emotional and social intelligence, describes people with self-regulation skills as being able to take responsibility for their actions, able to adapt to change, and the ability to respond appropriately to other people's irrational emotions or behavior.[241]

Self-regulation is the deepest level of group diagnosis. While easy to recognize, an absence of self-regulation may be the most difficult to correct quickly.

The lack of self-regulation stems from behaviors learned in childhood and reflects long-standing culture, values, personal experience, and beliefs. In a group setting, an inability to self-regulate impacts an individual's influence on decisions, ability to learn from others, and to be welcome in the group. If not addressed quickly, the group loses cohesion and productivity.

Use the following questions to diagnose self-regulation problems.

Is anyone disengaged, neither contributing nor responding?

Shy or dissatisfied individuals will withdraw from group activities no longer functioning as a group member.

Is anyone avoiding difficult conversations?

Individuals participate when they feel safe. They will avoid contributing or responding when at risk of saying something wrong in a sensitive situation.

240 National Research Council. 2011. Assessing 21st Century Skills: Summary of a Workshop. J.A. Koenig, Rapporteur. Committee on the Assessment of 21st Century Skills. Board on Testing and Assessment, Division of Behavioral and Social Sciences and Education. Washington, DC. The National Academy Press.

241 Daniel Goleman. 2005. *Emotional Intelligence: Why it Can Matter More Than IQ*. Bantam Books. New York, New York.

Does anyone seem to be holding back or softening their contributions?

Another way of avoiding risky situations is to speak so as not to offend anyone. This keeps the group from benefitting from the full value the individual was expected to contribute.

Is anyone exaggerating or being too forceful?

Individuals will exaggerate or be too forceful when they do not believe another is properly valuing their contribution or is valuing an alternative contribution too highly.

Is anyone interrupting or changing the subject?

Interrupting and changing the subject is an aggressive way to avoid a risky subject or an attempt to distract from a perspective the individual does not value.

Is anyone criticizing or attacking others?

Criticism and attacks indicate a significant self-regulation breakdown. The individual is frustrated that the group is undervaluing their contributions, they are undervaluing other's contributions, or they have a personal dislike for someone. Worst case, the individual feels ostracized by the group.

Breakdowns in self-regulation must be addressed swiftly. Return to group expectations. Do not single out an individual in front of the group.

Addressing self-regulation issues can require considerable effort outside group meetings. Reflect on empathy, compassion, tolerance, and patience before meeting with the individual. Assure them that they are valued by you and the group. Be specific about why they were asked to participate in the group. State that you want to help them convey their value in the most appropriate way so the group can make the best decision possible for the organization.

Use the path to action in the de-escalate stressful situations chapter to frame a conversation. Ask them to consider alternative perspectives without bias. Use the active listening techniques of the communicate effectively chapter. Help them monitor their emotional states, name the emotions, and decide if they are appropriate for the situation. Reframe negativity in a positive manner.

To model integrity, be clear about the consequences of behavior that diminish group cohesion and productivity. For example, inappropriate behavior will result in dismissal from the group and/or organization.

Managing group dynamics is a critical leadership skill. It is also one of the most difficult skills in which to achieve proficiency. Every group is a *de novo* situation. Be patient and trust that your integrity, desire to help others, and patience will allow you do the best you can.

IDENTIFY AND ANALYZE STAKEHOLDERS

All stakeholders must participate in the gains and losses of any particular situation.

Christine Lagarde, president of the European Central Bank and former chair of the International Monetary Fund

Stakeholders are individuals or organizations that have an investment, share, or interest in something. In the context of healthy lands and waters, stakeholders range from a property owner, to an individual tasked with managing a resource, to anyone dependent on the services provided by healthy lands and waters.[242]

Stakeholders are directly or indirectly invested in ecosystem services that provide food, water, raw materials, medicines, energy, regulate the climate and decompose waste, and enable recreation, education, and cultural practices. We are all stakeholders when the subject is healthy lands and waters.

Achieving conservation impact requires not only understanding biophysical factors that regulate species and ecosystems, but social factors like culture, economics, and politics. One method for improving your success in socioecological systems is by identifying stakeholders and analyzing their attributes. Who is interested or affected? What are their motivations, values, priorities, loyalties, fears, and hopes? Are there relationships among stakeholders? How

242 Millennium Ecosystem Assessment. 2005. Ecosystems and Human Well-being: Synthesis. Island Press, Washington, DC.

do these attributes inform planning and action? Can you develop empathy with your stakeholders?

Why do you care about stakeholders? Keeping lands and waters healthy is an adaptive and increasingly complex ecological, economic, political, and social challenge. None of us possess the depth and breadth of information, intelligence, and authority to succeed without stakeholders.

You need to work with others - biological, soil, and atmospheric scientists, educators, lawmakers and law enforcers, farmers, foresters, and fishers, and consumers of natural resource products and services.

But if everyone is a stakeholder, how do you work with everyone? You don't. You define a challenge or opportunity, identify the critical stakeholders for a situation, select a set of applicable attributes, qualify those attributes the best you can, and then analyze the system to determine potential solutions. Then, and only then, invite the stakeholders to participate in a collaborative and cooperative group with a goal of meeting the challenge or realizing the opportunity.

The Power of Visual Ideation

Visual ideating can help you identify and analyze stakeholders. Ideating is to form an idea and visual ideating capturing the idea in an image. Think Leonardo da Vinci.[243] He took notes, sketched inventions, and drew images in his notebooks. The combination of words and images enabled a quick understanding of his meaning.

Brain studies reveal that you process and recall visual images more strongly than written ones.[244,245,246,247] Your brain likes pictures, and drawing and connecting images trigger the more creative regions of the brain.

243 Walter Isaacson. 2018. Leonardo Da Vinci. Simon and Schuster. New York, New York.

244 Howard Levie and Richard Lentz. 1982. Effects of text illustrations: A review of the research. Educational Communication and Technology 30(4):195–232.

245 Stephen Kosslyn. 2007. Clear and to the Point: 8 Psychological Principles for Compelling PowerPoint Presentations. Oxford University Press. New York, New York.

246 Allan Pavio. 1986. Mental Representations: A Dual Coding Approach. Oxford University Press. New York, New York.

247 Janne Seppaanen. 2006. The Power of Gaze: An Introduction to Visual Literacy. Peter Lang, International Academic Publishers. New York.

The simultaneous depiction of ideas promotes working memory, the part of short-term memory that is concerned with immediate, conscious, perceptual, and linguistic processing.[248] By keeping information in short-term storage, the prefrontal cortex can manipulate and analyze it.

Working with information streaming in from other cortical areas, the prefrontal cortex can determine what information, if any, is relevant to the problem you are trying to solve. Once this overlapping of ideas occurs, cortical cells start to form new connections wiring themselves into new networks.

Identifying and Analyzing Stakeholders

The following approach to identifying and analyzing stakeholders is influenced by Cambridge Leadership Associates, systems analysis, and several mind mapping techniques.[249,250,251] The approach has four steps.

1. Define a challenge or opportunity.
2. Identify stakeholders.
3. Define stakeholder attributes.
4. Analyze the system.

Define a Challenge or Opportunity

Identifying and analyzing stakeholders begins by selecting a challenge or opportunity. A challenge should be a stimulating task or problem that requires others to resolve. An opportunity is a good chance for advancement or

248 Akira Miyake and Priti Shah (eds.). 1999. Models of Working Memory: Mechanisms of Active Maintenance and Executive Control. Cambridge University Press.

249 Ronald Heifetz, Alexander Grashow, Marty Linsky. 2009. The Practice of Adaptive Leadership. Harvard Business Review Press, Boston, Massachusetts, U.S.

250 Michael Powell. 2014. Productivity hacks: Da Vinci's mind mapping. 21 January. https://www.linkedin.com/pulse/20140121122453-271580474-productivity-hacks-davinci-s-mind-mapping/.

251 Matt Tanguay. No date (accessed 26 December 2019). How to mind map: Visualize your cluttered thoughts in 3 simple steps. https://www.lifehack.org/articles/work/how-to-mind-map-in-three-small-steps.html.

progress. To some extent, whether your situation is a challenge or an opportunity depends on your perspective. I usually speak to the challenge to create urgency, and the opportunity to provide hope.

Alternatively, you could replace your challenge or opportunity with a goal. The disadvantage of beginning with a goal is that stakeholders feel that the decision has already been made. Consequently, they may believe you are more interested in going through a process than meaningfully engaging them. The latter is more likely to ensure you have sustainable partners.

To begin, concisely and understandably write your challenge or opportunity in the middle of your work area.

Identify Stakeholders

Stakeholders are individuals or organizations that have an investment, share, or interest in something. Depending on the scale you are working at, the stakeholders could be individuals' names or descriptors for professional or social groups. To identify stakeholders, ask yourself a series of questions.

- What lands and waters could be encompassed by potential solutions?
- Who owns the land and water?
- Who has regulatory authority over the land and water?
- Who manages the land and water?
- Who relies on the land and water for their well-being?
- Who relies on the land and water for their prosperity?
- Who might be an advocate for a potential solution?
- Who might be an antagonist to a potential solution?

Connect the stakeholders to the central challenge or opportunity.

Define Stakeholder Attributes

Before you can engage stakeholders, you need to understand them and connect with them as individuals and as a group. What are they like? What keeps them up at night? What do you want them to do? How might they resist? Where are their loyalties? What motivates them?

You can apply many attributes to your stakeholders depending on your situation. Here are a few.

- What is the stakeholder's desired outcome?
- What is the stakeholder's risk?
- What is the stakeholder's opportunity?
- What are the stakeholder's power and influence over the outcome?
- What is the stakeholder's value or belief guiding their behavior and decision-making?
- What obligations does the stakeholder have to people outside their immediate group (e.g., customers, suppliers)?
- Does the stakeholder have explicit or hidden alliances with other stakeholders?

Each stakeholder is assigned the same set of attributes. To the best of your ability, qualify each attribute for each stakeholder. Apply empathy to adopt the stakeholder's perspective. Review websites for mission, vision, and actions.

If you have a good relationship with a stakeholder interview them but be circumspect about your plan to convene a group around a specific challenge or opportunity. Revealing your intent will bias the responses and lead to mistrust and disgruntlement before you can assemble a group.

An alternative to speaking directly with a stakeholder is to talk to someone you trust who has an existing relationship with the stakeholder. Use the active listening practice in the effective communication chapter and distinguish facts from stories with whomever you interview.

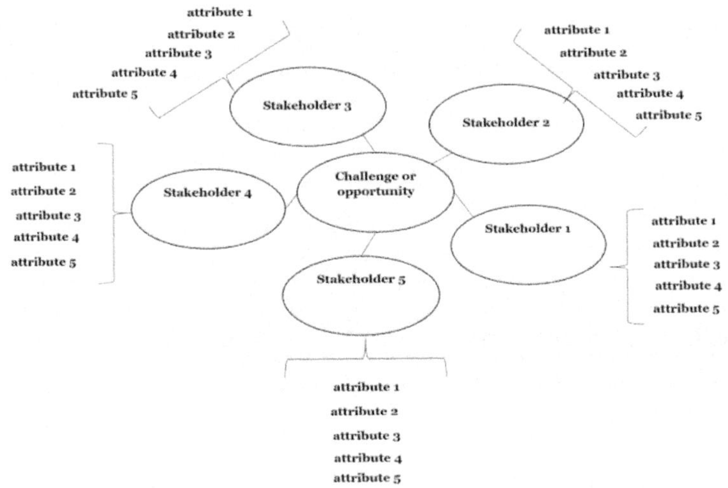

Analyze the System

Your goal is to analyze the stakeholder system in preparation for assembling and managing a group. Review and compare the attributes.

- Do the stakeholders have different goals?

- Are influence and power balanced or do some stakeholders have more than others?
- Are some stakeholders relatively powerless?
- Do the stakeholders vary in their level of interest?
- Will some stakeholders be very engaged, and others don't seem to care about the outcome?
- What are the risks to the stakeholders?
- Are there different risks or are they similar?
- What is the opportunity for the stakeholders?
- Are there different opportunities or are they similar?

To understand the system more deeply, you can examine some of the attributes individually or in combination. For example, a concentric circles diagram can help visualize the relative influence or power of the stakeholders. Interest, risk, and opportunity also lend themselves to this type of relative analysis.

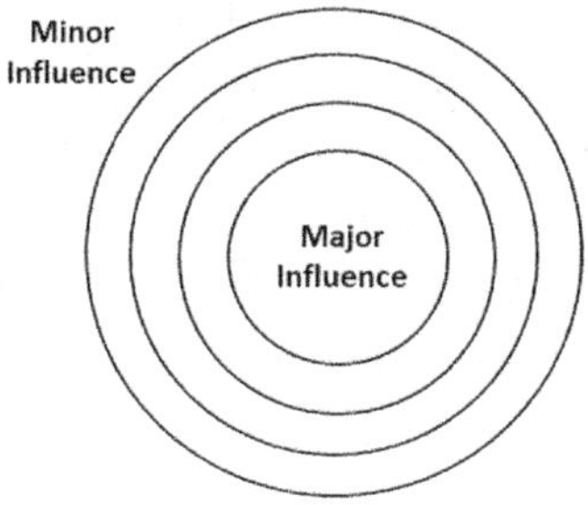

Stakeholder goals can be analyzed by depicting relative similarity across two or more dimensions. In the following example, two pairs of stakeholders have differing goals and the other stakeholder seems to have a compromise goal.

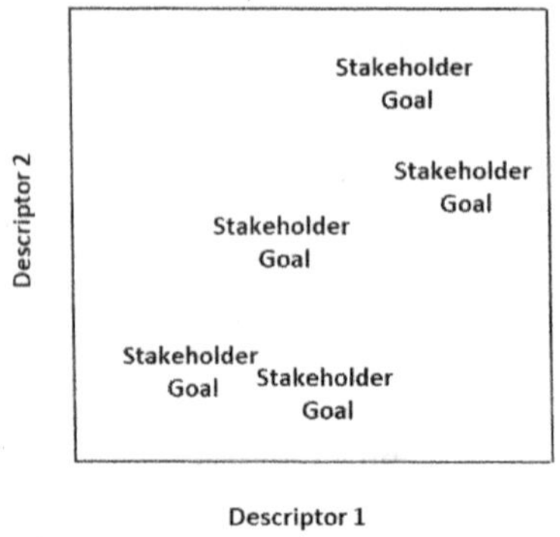

The analyses help you understand the complexity of the system needed to solve the challenge or realize the opportunity. Another benefit is a preliminary understanding of who will likely be trusting and supportive or mistrustful and opposed. Understanding both aspects will help you manage the stakeholder invitations and group function.

Your best option for accessing the full suite of alternatives is to avoid presuming an outcome. Groups are smarter than individuals and diverse groups are likely to discover solutions not previously recognized. If you have a preferred alternative, you can consider it during your analysis but keep an open mind when the group assembles. Group cohesion and productivity suffer if the members sense you have an agenda.

Your analysis is not intended for sharing with the group. The analysis is dynamic, and you will modify the analysis throughout the stakeholder meeting process to account for inaccurate and shifting attributes.

Lastly, your stakeholders should receive individual invitations that thoroughly address group structure. Schedule a face-to-face meeting to ensure effective communication including dispelling any reticence. The practices in the communication, de-escalating stress, and group dynamics chapters will help with the invitations and group function. Look for shared experiences, values, and goals to build connections.

Example

Here is the stakeholder identification and analysis process applied to conserving lakeshore endangered plant communities

Challenge

How do we conserve global and state rare plant communities on a state-owned lakeside natural area subject to long-standing and increasingly intense recreational use?

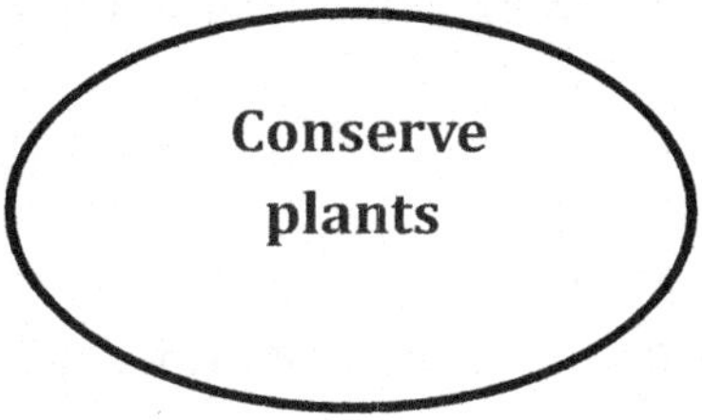

The natural area encompasses 400 acres near the mouth of a river on the south shore of a lake. The shoreline portion of the property is a sandy pond shore system comprised of four rare and imperiled natural communities. Two are globally rare, and the lake is the only location in the state where the communities are found. The pond shore system formerly supported numerous rare plant populations but currently only two are present. Impacts to the plant communities and species are from boater and local resident use of the beach-like setting.

Stakeholders

Boating Community

The boating community consists primarily of seasonal residents. Boaters have been using the beautiful white sand beaches of the natural area for more than 50 years. They are distraught about losing access to the beaches for the sake of a few plants.

Town

Elected officials want the state to provide a town beach for its residents within the natural area.

Lake Alliance

The Lake Alliance is a full-time volunteer organization dedicated to preserving and protecting the lake and its bays, rivers, and surrounding land as a unique recreational, environmental, and economic asset. The Alliance brings together individuals, property owner associations, and businesses in a broad community of interest to promote the careful use and shared enjoyment of the lake.

Lake Residents Association

Both seasonal and full-time resident priorities include peace and quiet, although they engage in various water-based activities. Concerns include noise, trash, and habitat destruction they attribute to the seasonal boating community.

State Agency Charged with Protecting the Resource

The agency finds, tracks, and facilitates the protection of the state's rare plants and exemplary natural communities. As mandated by statute, the agency determines protective measures and requirements necessary for the survival of native plant species in the state, investigates the condition and degree of rarity of plant species, and distributes information regarding the condition and protection of these species and their habitats.

The agency's rangers are authorized to police the natural area. However, they lack manpower and a boat needed to access the area.

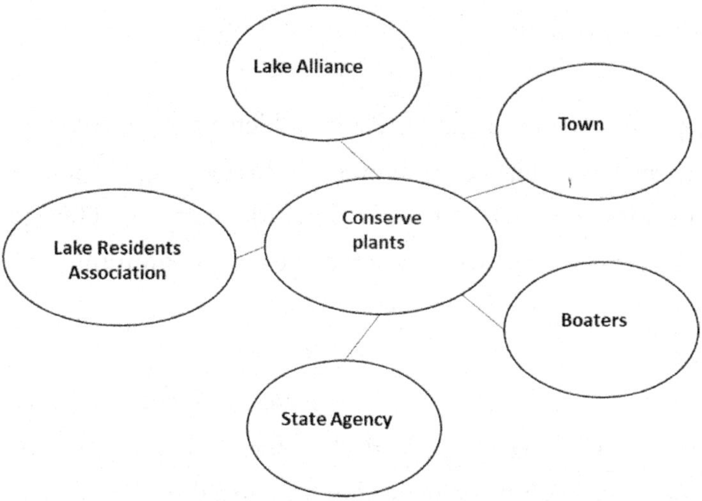

Defining Stakeholder Attributes

The stakeholder's attributes before convening are depicted in the following figure. Attributes include stakeholder goal, degree of influence, level of interest, risk, and opportunity.

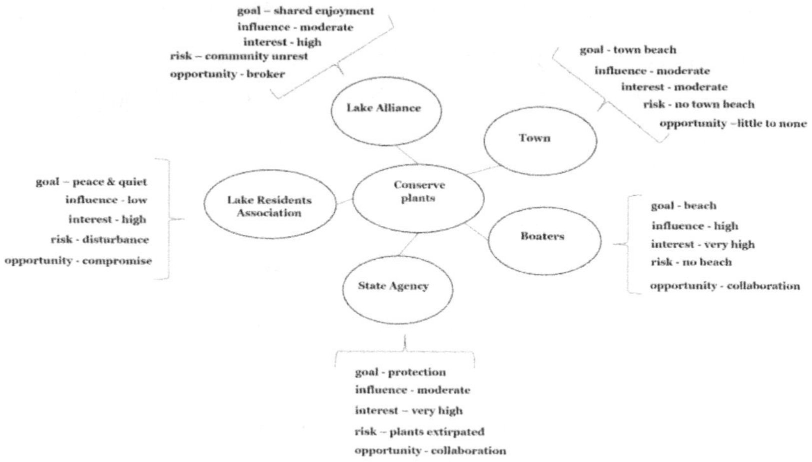

Analyze the System

Goal

Except for the state agency, the stakeholders want a recreational beach. In addition, the Lake Alliance wants everyone to get along and the Residents Association wants the boaters to tone down the partying. The agency is obligated under state statute to protect the rare plant communities and species.

Influence

The boating community had the greatest influence on the outcome because the state was unable to effectively enforce restrictions on use. The boating community was heavily invested in the natural area having used the beach section for 50 years. The number of boats had swelled to 200 or more on weekends. And many of the boaters had bought or built lakeshore weekend houses.

The town, Lake Alliance, and the agency had moderate influence over the outcome. The town's state representative had found common ground with colleagues that the state was interfering with local decision-making. The Lake Alliance members included local businesses dependent on recreational visitors, including the boaters. The agency should have had more influence, but state elected officials were not disposed to an authoritarian decision. The Lake Resident Association had few members, many of whom were apolitical.

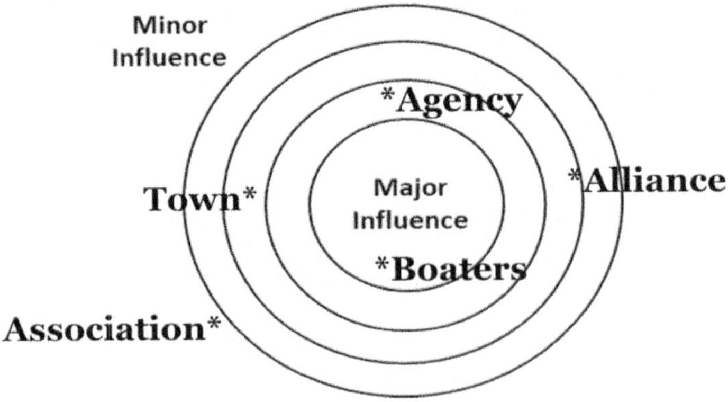

The difference in influence is evident in this diagram. The boaters have the most influence and the association the least influence. Other stakeholders have moderate influence. Any solution would require the cooperation of the boaters.

Interest

The boating community and the agency had extremely high interest in the outcome. The former for their legacy and investment and the latter for their obligation to fulfill statutory mandate. The resident association and the Lake Alliance had high interest.

The resident association was close to the beach and regularly complained about the boaters' noise and monopoly of the beach. The Lake Alliance wanted to protect their rural setting, particularly from excessive noise from boaters.

The town had moderate interest. They recognized that a town beach required parking, a boardwalk to protect rare plant communities, and lifeguards. On balance, few town officials were willing to invest if the state was not going to cover the costs.

Risk

The state agency was fearful the rare plant communities and species would be extirpated. The town and the boaters were concerned that the beach would be closed to recreation. The resident association was fearful the boaters would continue to disturb their peace and the Lake Alliance was concerned their group would fracture.

Opportunity

The town understood it had little opportunity given the sensitivity of the area and the cost of establishing and operating a town beach. The Lake Alliance recognized they had an opportunity to broker an agreement. The resident association, the boaters, and the state agency recognized an opportunity to compromise and collaborate.

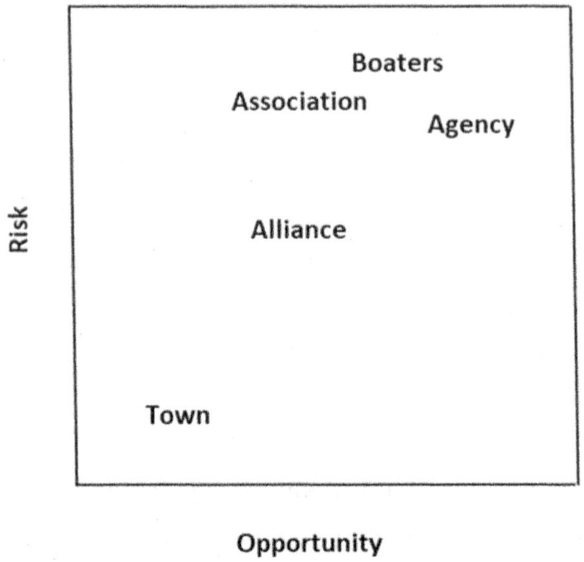

The figure indicates that the boaters and the agency have the most to lose and the most to gain. The Resident Association risks losing access to the beach, but has a moderate opportunity to work with the boaters and agency to reach a compromise. The Alliance is moderately at risk of alienating its members if it cannot help to find a solution and has a moderate opportunity to demonstrate value if it can help find a solution. The town has little opportunity to create a town beach and little risk if they do not.

Outcome

Under the leadership of the state agency, a stakeholder group formed to find a solution to the situation. Members included representatives from each of the stakeholder groups. The first meeting encompassed an icebreaker and discussion about purpose, goal, and expectations. The second meeting was used to evaluate proposed solutions. And the third meeting worked out the details of a preferred solution.

The state agency reached agreement with the boaters and the residents association for use of a portion of the beach in return for self-policing and help maintaining fencing and signage protecting sensitive areas. The boaters and the residents' association reached a noise abatement agreement. The solution has held for more than 10 years and the rare plant communities and species are extant, more robust, and expanding.

PART 5
CREATE AND INNOVATE

THE CREATIVE PROCESS

Imagination is more important than intelligence. For knowledge is limited to all we know and understand, while imagination embraces the entire world, and all there ever will be to know and understand.

Albert Einstein, genius, father of the theory of relativity, and influential science philosopher

Imagination is the visual component of creativity - the thinking process that helps you generate ideas. Creativity is your ability to transcend traditional ideas, rules, patterns, and relationships to create meaningful new ideas.

Creative people are better at recognizing relationships, making associations and connections, and seeing things in an original way. Seeing things that others cannot see.[252] Creativity is fulfilling and a central source of meaning.[253]

Past prosperity was due to an abundance of lawyers, doctors, accountants, engineers, and executives. Management guru Peter Drucker referred to these professionals as knowledge workers - people who get paid for putting to work what they learned in school.[254] Knowledge workers were educated to follow a set of established instructions down a single pathway to one conclusion. However, only 30 percent of job growth in the United States comes from knowledge work.

252 Nancy Andreasen. 2014. Secrets of the creative brain. July/August. The Atlantic. https://www.theatlantic.com/magazine/archive/2014/07/secrets-of-the-creative-brain/372299/.

253 Mihaly Csikszentmihalyi. 1996. The creative personality. Psychology Today. July. https://www.psychologytoday.com/us/articles/199607/the-creative-personality.

254 Peter Drucker. 1999. Knowledge-Worker Productivity: The Biggest Challenge. California Management Review 41(2):79–94.

Creativity is no longer reserved for artists. You live in an environment that is volatile, uncertain, complex, and ambiguous. An environment that requires new strategies and a new way of thinking, communicating, and creating. You must prepare yourself and those you lead for jobs that do not yet exist, using technologies that have not been invented, to solve problems you have not yet recognized.

Seventy percent of job growth comes from heuristic work and leading organizations are shifting their focus to heuristically skilled workers. [255]

- Workers who have the capacity to detect patterns and opportunities.
- Workers who can synthesize information and craft satisfying narratives.
- Workers who can combine seemingly unrelated ideas into something new.
- Workers who are creative and produce innovations.

Sustaining healthy lands and waters is complex, requiring science, policy, implementation, and adaptive management in response to dynamic stressors. You face a knowledge burden requiring lengthy education or narrowing expertise, both of which reduce individual innovative capacity.[256]

Creativity is a critically important leadership quality for success.[257] As a creative leader, you can foster conditions that can help to inspire creativity in others.[258] But first, you must understand the cognitive requirements for creative problem-solving.[259]

255 Daniel Pink. 2006. A Whole New Mind. Riverhead Books (Penguin Group). New York.

256 Benjamin Jones. 2008. The Burden of Knowledge and the 'Death of the Renaissance Man': Is Innovation Getting Harder? Working Paper 11360. National Bureau of Economic Research. Cambridge, Massachusetts.

257 Ladan Nikravan. 2012. Why creativity is the most important leadership quality. 30 May. https://www.chieflearningofficer.com/2012/05/30/why-creativity-is-the-most-important-leadership-quality/.

258 Louise Stoll and Julie Temperley. 2009. Creative Leadership: A Challenge of Our Times. School Leadership & Management 29(1): 65–78.

259 Roni Reiter-Palmon and Jody Illies. 2004. Leadership and creativity: Understanding leadership from a creative problem-solving perspective. Psychology Faculty Publications. 31.

The Role of the Brain in Creativity

As introduced in the decision making chapter the brain can be divided into four primary parts - brain stem, cerebellum, limbic system, and cerebral cortex. The brain stem is the simplest part of the brain and is responsible for heartbeat, blood pressure, and breathing. The cerebellum is evolutionarily newer than the brain stem, and controls balance, posture, and coordination.

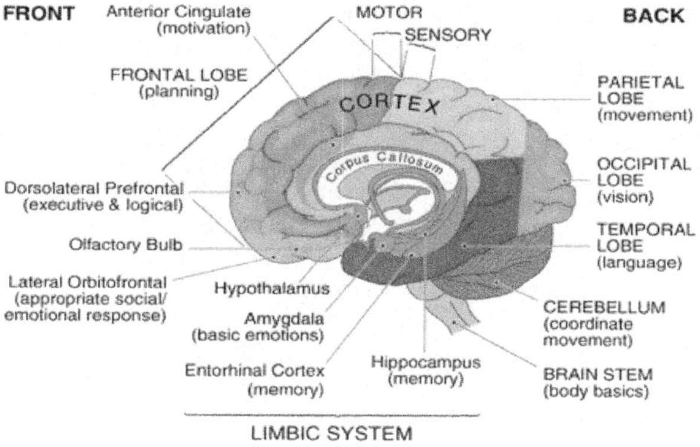

The two parts of the brain involved in creativity are the limbic system (aka midbrain) and the cerebral cortex. The limbic system encompasses several functionally and anatomically interconnected areas - the amygdala, hippocampus, hypothalamus, and thalamus - necessary for self-preservation. Conceptualized as the feeling and reacting brain, the limbic system regulates autonomic and endocrine function in response to stimuli.

The limbic system has been refined by evolution over the last several hundred million years, subjected to endless tests, so that it can make fast decisions based on scant information. Important decisions about complex challenges are best left to the limbic system.

The cerebral cortex or cerebrum is the largest part of the brain and is responsible for most of the brain's function. The cerebral cortex has been described as the thinking brain. A relatively recent development, the cerebral cortex is subject to errors and performs best with simple problems, decisions that do not matter too much, and novel challenges.

Structurally, the cerebral cortex has four lobes, the temporal, occipital, parietal, and frontal. The temporal lobe controls visual and auditory memories, the occipital lobe helps control vision, and the parietal lobe focuses on comprehension.

The frontal lobe controls intellect, judgment, behavior, attention, physical reactions, muscle movements, coordinated movements, smell, and personality. Important to creativity, the frontal lobe also controls creative thought, problem-solving, and abstract thinking.

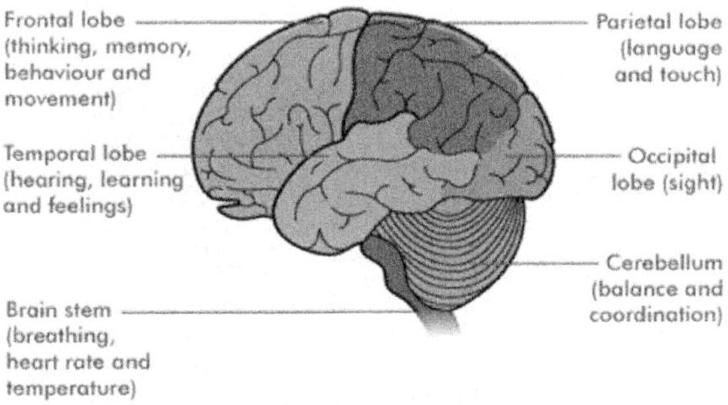

The two hemispheres of the cerebral cortex have different capacities. The left hemisphere of the brain is sequential, logical, literal, and analytical. The right hemisphere is nonlinear, intuitive, holistic, and inventive. The left hemisphere produces finished products. The right produces rough drafts.

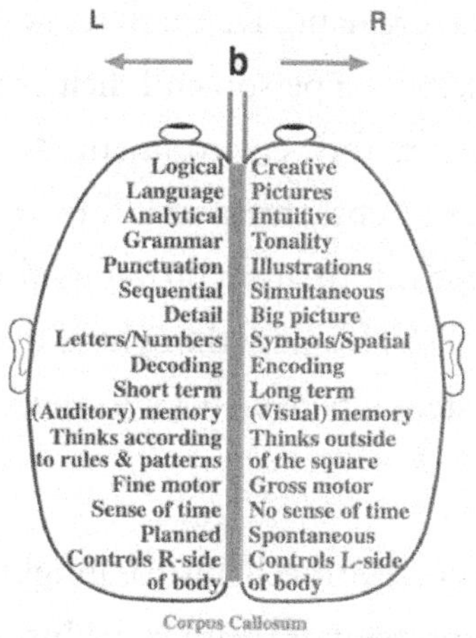

To understand how the brain can create, you must appreciate the enormous neural network you possess. Roughly 100 billion neurons capable of making 100 trillion distinct and elaborate connections with each other. Our neural network is plastic - capable of adopting new configurations. New ideas cause new assemblages of neurons to come together.

The brain has two fundamental creative mechanisms – working memory and insight.

Working Memory

Working memory is the less creative of the two mechanisms. It takes advantage of the dopamine reward pathway - a connection between the limbic system and the prefrontal cortex.

When the limbic system receives what it interprets as rewarding information, dopamine neurons (dopamine produces feelings of happiness) send the information to the prefrontal cortex, a mass of tissue behind the forehead that controls attention and allows us to focus. In the prefrontal cortex, the information enters working memory.

Working memory is where you temporarily store information you are using. Working memory can only hold a finite amount of information and is bombarded constantly by new thoughts and sensory experiences, all trying to distract you from the task at hand. Research suggests a close link between the working memory capacity of a person and their ability to selectively enhance or ignore information from their environment.[260]

In working memory, recently received information can be combined with other information. If the information proves useful the idea survives as a persistent link between cells. This new thought is then transmitted back to its source in the limbic system where neurons learn from the new idea.

Insight

The brain's second creative mechanism is insight. Insight is the mechanism most linked to an *aha* moment of creativity.[261] Insights are triggered by a piece of information that unexpectedly shifts your viewpoint or set of beliefs in a way that is transformative.[262] A new and penetrating understanding that allows you to see a problem or situation in a different light.

Every insightful creative journey begins with a bypass of your limbic system. Either because your limbic system doesn't recognize the challenge, or you are too stubborn to trust your instincts. The challenge provokes an intense mental search of your left cerebral hemisphere typically resulting in fatigue and frustration. Your dorsolateral prefrontal cortex, associated with impulse control, is deactivated.

The right hemisphere of your cerebral cortex then becomes active, searching for the subtle connections between seemingly unrelated information. Seeing not just the parts but how they come together to make sense of the whole.

An iterative formulation of possible solutions associated with smaller, partial insights precedes a moment of grander insight associated with your anterior superior gyrus, a small fold of tissue located on the surface of the right

260 Akira Miyake and Priti Shah (Eds.) 1999. Models of Working Memory: Mechanisms of Active Maintenance and Executive Control. Cambridge University Press.

261 Edward Bowden and Mark Jung-Beeman. 2003. Aha! Insight experience correlates with solution activation in the right hemisphere. Psychonomic Bulletin & Review 10(3):730–737.

262 Gary Klein. 2013. Seeing What Others Don't: The Remarkable Ways We Gain Insight. Perseus Books Group. New York, New York.

hemisphere just above the ear.[263,264] Your insight accompanies a sudden burst of gamma wave brain activity associated with the binding of neurons across the cerebral cortex forming a new network. Voila! Challenge solved.

The likelihood of producing an insightful solution is related to your resting-state brain activity - the ongoing acquisition and maintenance of information for interpreting, responding to, and predicting environmental demands. Your resting-state brain activity is modulated by mood and attention.[265,266] The time leading up to your moment of insight can be as important as the moment itself. And the moment itself is accompanied by a positive feeling of things coming together.[267]

Incidentally, an electroencephalogram can predict your moment of insight eight seconds before the insight consciously arrives.[268] The predictive signal is a steady rhythm of alpha waves emanating from the right hemisphere of your cerebral cortex. Alpha waves are closely related to relaxing activities (e.g., a warm shower). Insufficient alpha wave activity precludes insights.

7 Steps to Creativity

Your ideas do not just happen like a stroke of lightning. Instead, they are built through a chain of events and hard work. Rowan Gibson studied the process of creative thinking by successful inventors like Archimedes, Einstein, and Edison. He perceived those big ideas were built through a series of steps.[269] Here is a seven step process adapted from Gibson.

263 Edward Bowden, M. Jung-Beeman, J. Fleck, and J. Kounios. 2005. New approaches to demystifying insight. Trends in Cognitive Science. 9:322–328.

264 Sebastien Helie and Ron Sun. 2010. Incubation, insight, and creative problem solving: a unified theory and a connectionist model. Psychological Review. 117:994–1024.

265 John Kounios et al. 2006. The prepared mind: neural activity prior to problem presentation predicts subsequent solution by sudden insight. Psychological Science. 17:882–890.

266 Karuna Subramaniam, John Kounios, Todd Parrish, and Mark Jung-Beeman. 2009. A brain mechanism for facilitation of insight by positive affect. Journal of Cognitive Neuroscience. 21:415–432.

267 Sascha Topolinski and Rolf Reber. 2010. Gaining insight into the "Aha" experience. Current Directions in Psychological Science. 19:402–405.

268 Jonah Lehrer. 2012. Imagine: How Creativity Works. Houghton Mifflin Harcourt. Boston, Massachusetts.

269 Rowan Gibson. 2015. The 4 Lenses of Innovation: A Power Tool for Creative Thinking. John Wiley & Sons. Hoboken, New Jersey.

Step 1 - Frame a specific challenge and focus on solving it.

Every process has a beginning, and the creative process begins with knowing what challenge you want to solve. Name your challenge. Define it as thoroughly as possible. View it from different perspectives including those of colleagues and friends.

Step 2 – Learn from the work of others by researching the subject.

The odds are that you are not the first person to recognize the challenge and want to solve it. Research who else might have worked or is working on the challenge. What perspective did they bring to the challenge? What solutions did they try? Did their progress teach you anything? Did their failures inform your perspective? Think broadly. Are there analogous challenges in different disciplines?

Step 3 - Immerse yourself in the problem and explore possible solutions.

Combine your thoughts with what you learned from others and start experimenting. Do not artificially set a limit to what you will try. Try, learn, try, and learn until you have tried everything you can think of - funding and time notwithstanding.

Step 4 - Reach a roadblock, step away, detach, let the problem incubate, and feel the creative frustration.

Now is the time to step away from the challenge. Do not set a time limit. Work on another challenge. Spend time with family and friends. Read a book. Do anything but think about the challenge. The key is to detach your conscious mind from the challenge and let it ruminate in your unconscious mind.

Step 5 - Come to an illuminating insight that fundamentally shifts your perspective.

This is the easy step, letting the insight come to you. While your conscious mind was focused on something else your unconscious mind was combining seemingly unrelated information. Until, aha! One of those combinations fundamentally shifts your perspective, shows potential, and is brought to the attention of your conscious mind.

Step 6 - Build the insight into a big idea - a new combination of thoughts.

Your insight is not a fully formed idea. An insight is more like a potential missing puzzle piece. Combine your insight with your ideas - your other pieces of the puzzle - to form a new idea. Is it promising? Is it worth trying?

Step 7 - Test and validate the new idea by trying to make it work.

Return to your workspace and test your new idea with your re-engaged conscious mind. Does it solve your challenge? Yes? Congratulations. No? Return to step 5 and patiently wait for another insight. If the challenge is urgent (e.g., people are suffering, your boss is screaming ...) you can return to step one. Add some new perspectives to your working group including some who will challenge your assumptions.

Effective Creative Teams

Psychologist Keith Sawyer studied jazz groups and theater ensembles to understand collaboration. He then extended his studies to Google, Cisco, IDEO, 3M, W. L. Gore, and other successful businesses. His conclusion? Creativity is always collaborative even if you are alone[270].

Small sparks accumulate over time with multiple dead ends and a reinterpretation of previous ideas. An invisible collaborative web that eventually produces ideas and innovations.

Sawyer identified seven key characteristics of effective creative teams.

1. *Innovation emerges over time.* Each participant contributes a small idea and innovations occur when organizations combine the right ideas in the right structure.

270 Keith Sawyer. 2007. Group Genius: The Creative Power of Collaboration. Basic Books. New York, New York.

2. *Successful collaborative teams practice deep listening.* Participants listen for new ideas while at the same time generating their own ideas.
3. *Team members build on their collaborator's ideas.* When teams practice deep listening each new idea is an extension of previous ideas.
4. *Only later does the meaning of each idea become clear.* Ideas are not fully important until they are consumed, reinterpreted, and applied by others.
5. *Surprising questions emerge.* The most transformative creativity results when the team reframes a problem or finds a new problem that no one noticed. The most creative groups excel at finding new problems.
6. *Innovation is inefficient.* Many ideas are impractical. Some are good ideas but need other ideas to turn them into innovations. And only a few ideas produce innovations.
7. *Innovation emerges from the bottom up.* The most innovative teams can spontaneously restructure themselves in response to unexpected shifts.

PROMOTE CREATIVITY

Creativity is intelligence having fun.

Albert Einstein, genius of the theory of relativity and influential philosopher of science

A Creative Mindset

Is creativity inherited or learned? The answer is both. According to a study of twins about 22 percent of the variance in creativity is influenced by genes.[271] Musical creativity is correlated with a cluster of genes related to brain plasticity.[272] The same study also found increased creativity correlated with genes that affect processing of the neurotransmitter serotonin. Elevated serotonin levels increase brain connectivity, a key for creativity.[273]

A genetic basis for creativity suggests some people have an inherent tendency for creativity. But a lot depends on how genes are expressed which is a function of your environment. You can have a genetic predisposition for creativity but not be creative.

Fortunately, creativity is a mindset that can be learned, and you don't have to be a genius to learn how to be creative. The research-supported threshold theory states that creativity is not affected by intelligence at or above an IQ of

271 Barbara Kerr (Ed.). 2009. Encyclopedia of Giftedness, Creativity, and Talent. Vol. 1. Sage. Los Angeles, California.

272 Liisa Ukkola-Vuoti et al. 2013. Genome-wide copy number variation analysis in extended families and unrelated individuals characterized for musical aptitude and creativity in music. PLOS ONE 8(2): 1–9.

273 Christopher Kraus et al. 2013. Gray matter and intrinsic network changes in the posterior cingulate cortex after selective serotonin reuptake inhibitor intake. World Neurosurgery. 84:236–244.

120.[274,275] An IQ of 120 is good, but not high (140), and certainly not genius (160). Above an IQ of 120 personality factors are a better predictor of creativity.

Convergent and Divergent Thinking

General systems scientist George Land concluded there are two types of thinking processes when it comes to creativity[276]. Convergent thinking is focused on coming up with the single, well-established answer to a problem. You judge ideas, criticize them, refine them, combine them, and improve them. All of which happens in your conscious thought. Think working memory and one right answer.

The other thinking process is divergent. You generate creative ideas by exploring many possible solutions, usually subconsciously. Divergent thinking typically occurs spontaneously, flowing freely in a nonlinear fashion. Think insight and the potential for many right answers.

Becoming Creative

Lou Mobley, Director of IBM's Executive Development School realized that IBM's success depended on teaching executives to think creatively rather than teaching them how to read financial reports. His teaching was built around six startling insights.[277]

1. Traditional teaching methods like reading, lecturing, testing, and memorization are worse than useless.
2. People do not learn to be creative, they become creative.
3. Becoming creative is unlearning, rather than learning.
4. The fastest way to become creative is to hang around creative people.

274 George Land and Beth Jarman. 1992. Breakpoint and Beyond: Mastering the Future Today. HarperBusiness. New York, New York.

275 Nancy Andreasen. 2014. Secrets of the creative brain. July/August. The Atlantic. https://www.theatlantic.com/magazine/archive/2014/07/secrets-of-the-creative-brain/372299/.

276 George Land and Beth Jarman. 1992. Breakpoint and Beyond: Mastering the Future Today. HarperBusiness. New York, New York.

277 Lou Mobley and Kate McKeown. 1989. Beyond IBM: The Inside Story. McGraw Hill. New York, New York.

5. Creativity is highly correlated with self-knowledge.
6. Give your students permission to be wrong.

Based on a review of more than 70 studies, a well-designed program to help others become creative encourages a person to learn, discover, understand, or solve problems on his or her own.[278]

Skill is developed by experimenting, evaluating possible answers or solutions, or by trial and error. Often by using realistic exercises appropriate to the domain at hand. For example, how many ways can you make lands and waters healthy? A key is pushing people beyond the obvious answers.

Unlearning Creativity

Unfortunately, creativity can be unlearned and often is. George Land and collaborator Beth Jarman applied a creativity test developed for NASA recruitment to children.[279] Ninety-eight percent of five-year-old children scored at the genius level. At10 years old, 30 percent of the children scored at the genius level. And at 15 years old, only 12 percent of the children scored at the genius level. The same test given to 280,000 adults found only 2 percent to be creative geniuses.

So, what happened? Essentially all young children are creative and almost all adults are not. Schools and parenting are taking divergent thinkers, apparently a default in children, and teaching them how to be exclusively convergent thinkers. Traditional teaching methods like reading, lecturing, testing, and memorization are counterproductive to creativity.[280]

Re-learning Creativity

George Land suggested you need to allow people to split their thinking processes into convergent and divergent providing time for both to make each

278 Ginamarie Scott, Lyle Leritz, and Michael D. Mumford. 2004. The effectiveness of creativity training: A quantitative review. Creativity Research Journal 16(4):361–388.

279 George Land and Beth Jarman. 1992. Breakpoint and Beyond: Mastering the Future Today. HarperBusiness. New York, New York.

280 Lou Mobley and Kate McKeown. 1989. Beyond IBM: The Inside Story. McGraw Hill. New York, New York.

of them more effective. The good news is that schools, parents, and businesses are recognizing the criticality of divergent thinking. Creating time for both convergent and divergent thinking in learning plans for children and professional development plans for adults.

If you want your child to retain their ability and desire to be creative, and your employees to innovate, encourage them to let their minds run free while they come up with ideas. Only afterward critically evaluate those ideas and start working on the ones you think are best.

Attributes of a Creative Person

Do you know a creative person when you see one? Are you a creative person? Creative people generate ideas. They transcend traditional ideas, rules, patterns, and relationships. They recognize relationships and make associations and connections.

George Land and Beth Jarman offer three core concepts for creating transformation and change.[281]

- *Power of Creative Growth*: Being truly open to the unique thinking that energizes major breaks with the past.
- *Power of Future Pull*: Being pulled by the possibilities of the future to create what has never been seen before.
- *Force of Connecting*: Building open, accepting, partnerships of understanding to co-create new solutions that transcend traditional thinking.

Research suggests that good, creative leaders consistently develop the following attributes.[282]

- Knowledge of leadership styles and perceptions.

281 George Land and Beth Jarman. 1992. Breakpoint and Beyond: Mastering the Future Today. HarperBusiness. New York, New York.

282 Victor Sohmen. 2015. Reflection on creative leadership. International Journal of Global Business 8(1):1–14.

- Understanding of different cultures.
- Individual and team motivations.
- Relationship skills.
- An ability to manage change.
- Multiple communication styles.
- Listening ability.
- Decision-making skills.
- Personal ethics.

Mihaly Csikszentmihalyi, one of the most distinguished and accomplished psychologists of our time, finds creative people to be antithetical.[283] Remarkable for their ability to adapt to almost any situation, creative people show tendencies of thought and action that in most people are segregated. Instead of being an individual, each of them is many people.

- A great deal of physical energy but often quiet and at rest.
- Smart yet naïve.
- A combination of playfulness and discipline or responsibility and irresponsibility.
- Alternating between imagination and fantasy and a rooted sense of reality.
- Extroverted and introverted.
- Humble and proud.
- Escape rigid gender role stereotyping.
- Rebellious and conservative.
- Deeply passionate about their work yet extremely objective.
- Openness and sensitivity expose them to suffering and pain yet also to a great deal of enjoyment.

283 Mihaly Csikszentmihalyi. 1996. The creative personality. Psychology Today. 1 July. https://www.psychologytoday.com/us/articles/199607/the-creative-personality.

Methods for Promoting Creativity

Creativity is an essential leadership skill. You may be genetically predisposed toward creativity or excelled at creativity as a child. Regardless, you likely unlearned creativity to succeed in school or your career.

Fortunately, you can rediscover your creativity and align it with your goal of healthy lands and waters. And as a leader, you can help others rediscover the creativity necessary to succeed individually and collectively.

Promoting Your Own Creativity

Be Happy

You are more creative when you feel happy, are intrinsically motivated by your work, and have positive perceptions of your colleagues.[284] Positive emotions open your heart and your mind making you more receptive and creative. Conversely, a negative mood makes it easier for you to focus on details.

Don't Think About It

You can think too much and choke on your thoughts.[285] The part of your brain that monitors behavior, a network centered in the prefrontal cortex, starts to interfere with decisions that you normally make without thinking. Your brain begins to second-guess itself about skills that have been honed through years of diligent practice.

By not thinking about a problem you allow your brain to enter a phase that assimilates new information and explores alternative strategies. Take a walk. Read something enjoyable. Go on vacation.

284 T. Amabile and S. Kramer. 2011. The Progress Principle: Using Small Wins to Ignite Joy, Engagement, and Creativity at Work. Harvard Business Review Press. Boston, Massachusetts.

285 Jonah Lehrer. 2009. How We Decide. Houghton Mifflin Harcourt. Boston, Massachusetts.

Relax

Forcing yourself to concentrate can inhibit creativity. Relaxing allows you to hear the quiet voice in the back of your head telling you the answer.[286]

Also, a sharp drop in visual cortex activity precedes insight. Close or cover your eyes.

Meditate

Gathering the mind meditation (discussed in the chapter on de-escalating stressful situations) focuses on the breath. You are aware of and note all the physical and mental sensations that arise without focusing on any single sensation. The technique promotes divergent thinking and the generation of many new ideas.[287] Contemplative meditation, focusing on a single thought or concept, does not promote either divergent or convergent thinking.

Daydream

Your brain is extremely busy during daydreaming which facilitates an elaborate electrical conversation between the front and back parts of your brain.[288] Instead of responding to the outside world your brain starts to explore its inner base, blending concepts that are normally filed away in different areas.

People who consistently engage in daydreaming score significantly higher on measures of creativity. However, it is not enough just to daydream. You must maintain enough awareness to interrupt yourself and note creative thoughts.

Sleep

During sleep your internal censor the dorsolateral prefrontal cortex becomes quiet. Neurons throughout the brain release acetylcholine which

286 Geoffrey James. 2015. Neuroscience: Relaxing makes you more creative. 12 January. https://www.inc.com/geoffrey-james/neuroscience-relaxing-makes-you-more-creative.html.

287 Lorenza Colzato, Ayca Ozturk, and Bernhard Hommel. 2012. Meditate to create: the impact of focused-attention and open-monitoring training on convergent and divergent thinking. Front. Psychol., 18 April 2012. https://www.frontiersin.org/articles/10.3389/fpsyg.2012.00116/full.

288 Jonah Lehrer. 2012. Imagine: How Creativity Works. Houghton Mifflin Harcourt. Boston, Massachusetts.

stimulates semi-random and unpredictable connections among neurons. You are exposed to an excess of surprising connections and strange ideas.[289]

Most will be useless but if you are lucky you will find answers in the middle of the night and remember them when you wake. Conversely, sleep deprivation diminishes creativity.[290,291] Try to get a good night's sleep every night.

Exercise

Anecdotes suggest that bodily movement helps overcome mental blocks and lack of inspiration. Several studies suggest that physical exercise may sometimes enhance creative thinking but the methods used render the evidence inconclusive.[292,293]

A more focused study found that exercise interfered with divergent thinking.[294] The impact on convergent thinking depended on the training level. Non-athlete (i.e., did not exercise on a regular basis) cognitive performance was impaired by exercise while athletes (i.e., exercised at least three times a week during the last two years) obtained some benefit.

Get Outdoors

Your busy, technology-laden life places high demands on the attention parts of your brain. Attention restoration theory suggests that nature can help restore mental fatigue and concentration and encourage more effortless brain function.[295]

289 Matthew Walker, Connor Liston, J. Allan Hobson, and Robert Stickgold. 2002. Cognitive flexibility across he sleep-wake cycle: REM-sleep enhancement of anagram problem solving. Cognitive Brain Research 14:317–324.

290 Yuntao Bai, Helen Guohong, and P. Harris. 2017. Nightmare bosses: The impact of abusive supervision on employees' sleep, emotions, and creativity. Journal of Business Ethics. 145(1). https://www.researchgate.net/publication/283575012_Nightmare_Bosses_The_Impact_of_Abusive_Supervision_on_Employees'_Sleep_Emotions_and_Creativity.

291 James Horne. 1988. Sleep loss and divergent thinking ability. December. Sleep 11(6):528–536.

292 Joan Gondola, and Bruce Tuckman. 1985. Effects of a systematic program of exercise on selected measures of creativity. Perceptual Motor Skills. 60:53–54.

293 Benjamin Baird et al. 2012. Inspired by distraction: mind wandering facilitates creative incubation. Psychological. Science. 23:1117–1122.

294 Lorenza Colzato et al. 2013. The impact of physical exercise on convergent and divergent thinking. Frontiers of Human Neuroscience. December. https://www.frontiersin.org/articles/10.3389/fnhum.2013.00824/full?utm_source=newsletter&utm_medium=web&utm_campaign=Psychology-w4-2014.

295 Stephen Kaplan. 1995. The restorative benefits of nature: Toward an integrative framework. Journal of Environmental Psychology 15(3):169–182.

Exposure to natural stimuli is emotionally positive and low arousing. To benefit, you must be immersed in the environment, be away from habitual activities, the environment must effortlessly capture your attention, and you must welcome the exposure.

One aspect of the restoration is enhanced creativity. For example, four days immersed in nature improved hikers' creativity and problem-solving by 50 percent.[296]

Self-learn and Cross Boundaries

Do not wait for someone to teach you, be proactive and teach yourself. Standard ways of learning and teaching are not always the best way to be creative and may inhibit creativity.

What do you want to know? What do you need to know? Build wide-ranging knowledge. Develop multiple areas of expertise. Speak different languages. Pursue a passion that has nothing to do with your profession.

Work Hard

You cannot always relax, daydream, and sleep while waiting for an insight. Sometimes the creative process requires persistence.[297] Staring at a problem until it makes sense.

Your ability to make separate ideas coexist in the conscious mind is a crucial creative tool. Fortunately, you can rely on working memory - keeping information in short-term storage where it can be consciously contemplated.

Connections made in working memory do not feel mysterious like an insight. Instead, creative thoughts tend to be minor and incremental and will likely require editing before producing a fully formed idea.

296 Ruth Ann Atchley, David Strayer, and Paul Atchley. 2012. Creativity in the wild: Improving creative reasoning through immersion in natural settings. PLoS One 7(12):35174. https://journals.plos.org/plosone/article?id=10.1371/journal.pone.0051474

297 J. Michael Crant. 1995. The proactive personality scale and objective job performance among real estate agents. Journal of Applied Psychology 80(4):531–537.

Learn from Mistakes

Being wrong forces you out of your comfortable assumptions, and makes you explore options you had not considered.[298] Treat your mistakes like opportunities to learn while remaining focused on your goal.

Be Adaptable

Mihaly Csikszentmihalyi, creator of the theory of flow, studied the lives of over 90 eminent creative producers and thinkers to learn what they had in common. What he found is that the creative personality has no template. Being adaptable is key to creating effectively so you cannot expect your personality to be the same from one creative venture to the next.

Be an Outsider

Your thinking is shackled by the familiar.[299] For the sake of efficiency your brain spends a lot of time choosing what not to notice.

Creativity tends to peak after a few years of work unless you adopt an outsider mindset. Periodically ask yourself, *If I were just starting this job, what would I do differently*?

Practice wishful thinking. Imagine what the ideal solution would be regardless of financial, physical, or even ethical constraints. Put yourself into new situations. Confront existing paradigms. Try to forget what you already know.

Find new colleagues or change your career. Travel, especially to other countries, is one of the best ways to cultivate an outsider's perspective. Your distance from challenges at home and exposure to a new culture enhance open-mindedness.

298 Steven Johnson. 2010. Where Good Ideas Come From: The Natural History of Innovation. Riverhead Books. New York, New York.

299 Scott Kaufman and Carolyn Gregoire. 2016. Wired to Create: Unraveling the Mysteries of the Creative Mind. Tarcher and Perigee. New York, New York.

Re-purpose Ideas

An important creative tool is making connections across seemingly unrelated questions, problems, or ideas.[300] You can adopt an inquisitive attitude toward others' opinions and insights.

Develop your knowledge across fields to promote conceptual blending. Force your mind to approach intellectual roadblocks from new angles. Borrow tools from one discipline to solve problems in another.

Encourage your cognitive overlap by having several projects going at once or reading several different books at one time. Allow your mind to move through several different boxes. Your act of creation is really the act of recombination.

Exercise Outsight

Outsight is your capacity to perceive external things. According to a study of global CEOs two of the three most significant sources of creative ideas are outside the organization.[301] Outsight helps you to break free of pre-existing views, re-categorize information, and move beyond your habits to imagine novel alternatives.

You can begin creativity-building exercises and idea generation efforts by seeking personal experiences that directly confront implicit or explicit assumptions.[302] Changing the business as usual environment requires staying in touch with the world around you.

Red versus Blue

The color of your environment can affect the way you think.[303] Red is conducive for skills that require accuracy and attention to detail. Conversely, blue is associated with poor performance at short-term memory tasks but is twice

300 C. Christensen, J. Dyer, and H. Gregersen. 2009. The innovator's DNA. Harvard Business Review 87(12):60–67.

301 Steven Johnson. 2010. Where Good Ideas Come From: The Natural History of Innovation. Riverhead Books. New York, New York.

302 Marla Capozzi, Renee Dye, and Amy Howe. 2011. Sparking creativity in teams. April McKinsey Quarterly.

303 Juliet Zhu. 2009. Effect Of Colors: Blue Boosts Creativity, While Red Enhances Attention To Detail. Science Daily. 6 February. https://www.sciencedaily.com/releases/2009/02/090205142143.htm.

as effective as red for tasks requiring imagination. Blue stimulates alpha waves which are associated with relaxation and are a prelude to insight.

Visually Ideate

Brain studies reveal that you process and recall visual images more strongly than written ones. The brain likes pictures.

Drawing and connecting images activates the more creative regions of the brain and can trigger insights. The simultaneous depiction of ideas can also promote the minor and incremental ideas produced by working memory. See the chapter on identifying and analyzing stakeholders to learn about visual ideating.

Promoting Group Creativity

Creative leaders foster conditions that can help others become creative.[304]

- Stimulate a sense of urgency when necessary.
- Expose colleagues to new thinking and experiences.
- Provide time and space to facilitate the practicalities.
- Set high expectations.
- Promote individual and collaborative creative thinking and design.
- Use failure as a learning opportunity.
- Relinquish control.
- Model creativity and risk-taking.

Here are several methods for promoting your group's creativity.

Embrace Collective Intelligence and Crowd Source

Creativity is always collaborative even when you are alone.[305] Many ideas come from collective, distributed processes with many groups working on the

304 Louise Stoll and Julie Temperley. 2009. Creative leadership: A challenge of our times. School Leadership & Management 29(1):65–78.

305 Keith Sawyer. 2017. Group Genius: The Creative Power of Collaboration. Basic Books. New York, New York.

same problem[306] Groups can allow their ideas to flow freely into an environment populated by information where connection is valued more than protecting secrets.

Higher performing groups have significantly more communication with people outside their units.[307] Collective team efforts reduce the pressure and burden on individuals. The free flow of ideas and access to the best brains enables innovation to thrive, and problems to have elegant solutions sooner rather than later.

Balance Group Intimacy

The most productive groups are a mixture of old and new acquaintances.[308] This ensures interactions are comfortable, but not so comfortable as to preclude the interjection of new ideas.

The most creative and innovative organizations ensure that talented people from diverse backgrounds work together. This can be accomplished by having everyone in the same building and creating gathering spaces within and outside the workspace for informal discussion.[309] The highest-performing employees, the ones with the most useful new ideas, are the ones who consistently engage in the most interactions.[310]

Create a Climate for Re-discovering Creativity

Great creative groups have trust, familiarity, and a shared commitment.[311] You can create a climate for rediscovering creativity by balancing formal organizational hierarchy with informal visits to colleagues and partners to identify synergies. Ask how they are doing and what is important to them. Create

306 Steven Johnson. 2010. Where Good Ideas Come From: The Natural History of Innovation. Riverhead Books. New York, New York.

307 Alex Pentland. 2010. To signal is human. American Scientist. 98:204–211.

308 Jonah Lehrer. 2012. Imagine: How Creativity Works. Houghton Mifflin Harcourt. Boston, Massachusetts.

309 Ed Catmull (with Amy Wallace). 2014. Creativity, Inc.: Overcoming the Unseen Forces That Stand in the Way of True Inspiration. Random House. New York, New York.

310 Thomas Allen. 1984. Managing the Flow of Technology. MIT Press. Cambridge, Massachusetts.

311 Keith Sawyer. 2017. Group Genius: The Creative Power of Collaboration. Basic Books. New York, New York.

a climate of inquiry, openness, patience, and encouragement to promote learning and nurture a growth mindset.

Assign meaningful work to people so that they can see how their efforts contribute significantly to outcomes. Give staff time to pursue speculative ideas. Make it safe for others to try, fail, and learn from their experiences. Offer helpful suggestions to people as they try to learn and bounce back from mistakes.

Manage Disequilibrium

Healthy lands and waters are complex challenges requiring creative solutions. Comfort is the enemy of creativity. You can promote creativity through discomfort. A useful framework is the productive zone of disequilibrium.[312]

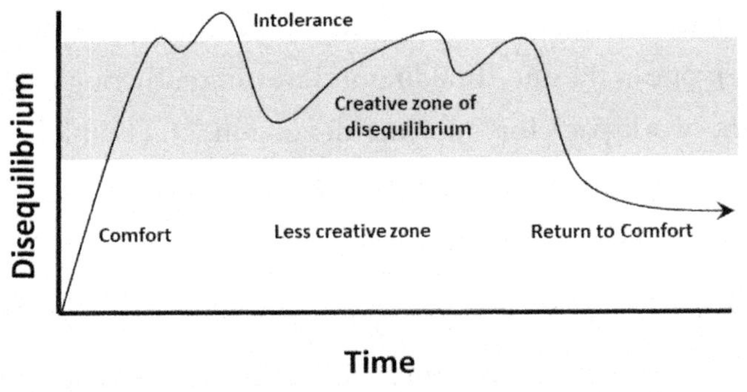

Source: Adapted from Ronald Heifetz et al. 2009.

A disequilibrium is a loss or lack of equilibrium or stability. Groups are comfortable at low disequilibrium. Their challenges have known solutions and the work is routine. When a new, complex challenge arises you need to stimulate their urgency and creativity. Make it more uncomfortable for them not to solve the challenge than to solve the challenge. But not so uncomfortable that they cannot think.

312 Ronald Heifetz, Alexander Grashow, and Marty Linsky. 2009. The Practice of Adaptive Leadership: Tools and Tactics for Changing Your Organization and the World. Harvard Business Review Press. Boston, Massachusetts.

Choose among the creativity promoting practices based on your group's strengths and weaknesses. Set a time limit. Take them for a walk outdoors. Ask them to think like an outsider, be adaptable, and exercise outsight. Have them visually ideate in subgroups and report back.

When the group pauses, encourage them to relax, meditate, daydream, sleep on it, and exercise. Keep them in the zone of creativity until a solution is found. And then congratulate them, thank them for their time and effort, and allow them to return to their comfort zone until the next challenge arises.

Brainstorming, Brainswarming, Criticism, and Plussing

Brainstorming, developed by advertising executive Alex Power in 1939, has been the single most popular technique for generating creativity.[313] The technique is losing its popularity, but at its core it consists of throwing out hundreds of ideas. Brainstorming takes several forms, but generally includes the following steps.

1. Prepare by setting a time limit, identifying a facilitator and recorder, and establishing ground rules.
2. Determine whether to freewheel, share ideas simultaneously or round robin with everyone taking a turn.
3. Generate as many ideas as possible. Unusual ideas are encouraged, and no criticism is allowed. Ideas can be combined and improved.
4. The ideas are filtered using a set of criteria

Brainstorming does not always release the potential of the group and can suppress it.[314] Brainstorming problems include fixating on one idea, making decisions as a group in a way that discourages creativity or individual responsibility and leads to conformity (i.e., groupthink), not allowing time for idea incubation, dominance by extroverts, and participants protecting good ideas.

313 Alex Osborn. 1948. Your Creative Power. Charles Scribner and Sons. New York, New York.

314 Keith Sawyer. 2017. Group Genius: The Creative Power of Collaboration. Basic Books. New York, New York.

Brainswarming is an improvement over brainstorming. Groups using brainswarming averaged 115 ideas in 15 minutes compared to 100 ideas in 60 minutes using brainstorming.[315]

Inspired by swarms of insects, brainswarming is a solo event that uses sticky notes or an electronic whiteboard.[316] A facilitator places a goal at the top and resources at the bottom. Group members individually add and build ideas working top down or bottom up. Potential solutions arise when the goal is connected to the resources.

For example, a completed brainswarming exercise for healthy lands and waters might look like the following.

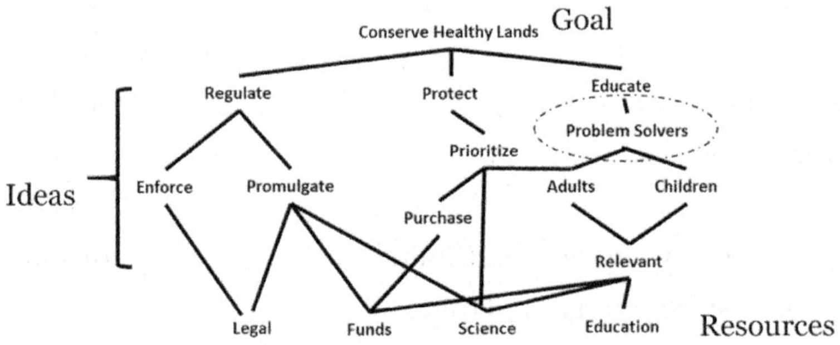

The goal of the exercise is to conserve healthy lands and waters. Available resources include legal, funds, science, and education. Possible solutions appear between the goal and the resources. In this example, the chosen solution is to create problem solvers.

Both brainstorming and brainswarming have their place. But the ideal creative idea-generating practice is a face-to-face, balanced practice. Research shows that criticism plays a productive role in driving creativity. Groups that critique each other produced 25 percent more ideas than groups asked not to

315 Tony McCaffrey. 2014 BrainSwarming: Because brainstorming doesn't work. https://hbr.org/video/3373616535001/brainswarming-because-brainstorming-doesnt-work.

316 Tony McCaffrey and Angelia Herrin. 2014. BrainSwarming: Because brainstorming doesn't work. Harvard Business Review. 24 June.

critique each other.[317] The practices in this book should help you lead a group's candid discussion of flaws characterized by debate and criticism.

You can use plussing to keep criticism from getting out of control.[318] Plussing requires that all criticisms be accompanied by a new idea that builds on perceived flaws in a productive manner. Mixing criticism with a new idea allows the group to immediately begin focusing on how to correct the flaw.

Plussing requires strong, adept leadership and enough participants with positive, emotionally intelligent relationships to model appropriate behavior for newcomers. The de-escalating stress and diagnosing group dynamics practices can help.

317 Charlan Nemeth, Bernard Personnaz, Marie Personnaz, and Jack Goncalo. 2014. The liberating role of conflict in group creativity: A study in two countries. European Journal of Social Psychology. 34:365–374.

318 Ed Catmull (with Amy Wallace). 2014. Creativity, Inc.: Overcoming the Unseen Forces That Stand in the Way of True Inspiration. Random House. New York, New York.

TURN IDEAS INTO INNOVATIONS

Innovation is seeing what everybody has seen and thinking what nobody has thought.

Albert Szent-Györgyi, Hungarian biochemist and 1937 Nobel Prize winner in Physiology or Medicine

The What and Why of Innovation

Creativity is the process of developing new ideas. Innovation is the process of implementing those ideas. An innovation can be a product, a process, or a way of managing or marketing.[319] Innovation finds and assigns value to creativity. Innovation is what solves a social need.

Innovation is a competitive advantage. Organizations mastering a majority of eight innovation essentials outperform others.[320,321] Innovative organizations:

Aspire

Innovative organizations regard innovation-led growth as critical and set targets. They signal leadership aspirations to staff through words, quantify the value of innovations, and include them in the organization's strategy.

319 Gerald Haman. 2000. Fostering creativity and innovation as a process: How to implement ideas that produce profitable solutions. Journal of Innovative Management. Summer, pp. 31–38.

320 Marc de Jong, Nathan Marston, and Erik Roth. 2015. The eight essentials of innovation. April. McKinsey Quarterly. https://www.mckinsey.com/business-functions/strategy-and-corporate-finance/our-insights/the-eight-essentials-of-innovation.

321 Daniel Cohen, Brian Quinn, and Erik Roth. 2019. The innovation commitment. October. McKinsey Quarterly. https://www.mckinsey.com/business-functions/strategy-and-corporate-finance/our-insights/the-innovation-commitment.

Choose

Innovative organizations invest in a coherent, balanced portfolio because of the uncertainty of predicting which innovations will be successful. They support several innovations but develop criteria for proceeding or terminating.

Discover

Innovative organizations seek actionable and diverse insights by scrutinizing which problems to solve, technology solutions, and effective business models. Discovery is iterative - prototyping helps learning by developing, testing, validating, and refining innovations.

Evolve

Innovative organizations create new business models that accommodate their innovations. And they are constantly re-evaluating the set of activities needed for the organization's innovations to perform.

Accelerate

Innovative organizations understand that good ideas arise in multiple organizations. Urgency is balanced against the risk of producing a flawed innovation. They achieve balance by testing ideas with stakeholders early in the process and establishing cross-functional teams.

Scale

Innovative organizations consider the magnitude and reach of their innovations. Findings define resource needs and capabilities and the risk associated with pursuing and launching an innovation.

Extend

Innovative organizations understand that innovations often require the skills and talents of external collaborators. They develop networks and regularly review members for relevancy. And they strive to be a partner of choice for others by being fair and transparent in dealings.

Mobilize

Innovative organizations ensure staff are motivated, rewarded, and organized to continuously innovate. They integrate innovation, strategy, and performance and promote collaboration, learning, and experimentation. Innovative organizations help people share ideas and ensure that teams have the right level of intimacy.

Do individuals and organizations want to innovate? Yes. Innovation solves problems, enables adaptation to change, and keeps them competitive in the marketplace. About 80 percent of organizations want to innovate but only 40 percent report being prepared to innovate.[322,323]

Given that innovation is a competitive advantage, why don't all organizations innovate? The answer is fear of change and risk of failure. Of every seven innovation projects about four enter development, one and a half are launched, and only one succeeds.[324]

Prepare Your Organization to Innovate

To innovate you need the right people and the right leadership. Begin with innovative people.[325,326,327] Innovators are passionately involved in what they are doing, they know their purpose, and they want to get to the problem and do something.

322 Nidhi Sahni, Laura Lanzerotti, Amira Bliss, and Daniel Pike. 2017. Is your nonprofit built for sustained innovation? August. The Bridgespan Group. https://www.bridgespan.org/insights/library/strategy-development/innovation-capacity-summary-and-resources.

323 Michael Ringel. 2015. Innovation in 2015. Boston Consulting Group. https://www.bcg.com/publications/2015/growth-lean-manufacturing-innovation-in-2015.aspx.

324 Gijs van Wulfen. 2014. 10 ways to innovation success in 2015. 23 December. https://www.linkedin.com/pulse/10-ways-innovation-success-2015-gijs-van-wulfen.

325 Carmine Gallo. 2014. The 7 innovation secrets of Steve Jobs. Forbes. 2 May. https://www.forbes.com/sites/carminegallo/2014/05/02/the-7-innovation-secrets-of-steve-jobs/#27482efb751c.

326 Hayagreeva Rao, Robert Sutton, and Allen Webb. 2008. Innovation lessons from Pixar: An interview with Oscar-winning director Brad Bird. April. McKinsey Quarterly, pp. 1–9.

327 Joe Robinson. 2014. The 7 traits of successful entrepreneurs. 10 January. Entrepreneur. https://www.entrepreneur.com/slideshow/299822.

Innovators have bold, specific visions that inspire others. They believe that the impossible is possible. Innovative organizations tolerate and embrace people that deviate from accepted standards, individuals who are restless and unconventional.

Innovators have broad knowledge acquired by associating with different fields and applying what they learn to what they are working on. And innovators create great user experiences. Users do not care about your innovation unless it enriches their life.

Innovative people need leadership. The Swedish Employment Council found 67 percent of innovation is determined by an organization's leadership.[328] Leaders promote innovation by establishing an inspiring vision and clarifying the goal, and committing to learning from continuous, intentional, systematic exploration, experimentation, trial, and error.[329,330]

Leaders create a safe environment that allows ideas to be expressed without fear of shame or blame[331]. An environment where everyone's opinion is valued and respected and feedback is honest and constructive. All staff can be visionaries and contribute to innovation.[332]

Structure your organization to support innovation. Include innovation in your strategy balanced with ongoing, revenue-producing initiatives.[333] Manage and reward good ideas, including those about how you operate. Eliminate barriers to innovation like no time to think, isolation, withheld information, and fear that failure will lead to termination.[334] Construct diverse teams with

328 Gerald Haman. 2000. Fostering creativity and innovation as a process: How to implement ideas that produce profitable solutions. Journal of Innovative Management. Summer, pp. 31–38.

329 Nidhi Sahni, Laura Lanzerotti, Amira Bliss, and Daniel Pike. 2017. Is your nonprofit built for sustained innovation? August. The Bridgespan Group. https://www.bridgespan.org/insights/library/strategy-development/innovation-capacity-summary-and-resources.

330 George Land. 2000. Creating a sustainable competitive edge. Journal of Innovative Management. Summer, pp. 3–10.

331 Brene Brown. 2012. Daring Greatly: How the Courage to be Vulnerable Transforms the Way We Live, Love, Parent, and Lead. Gotham Books. New York, New York.

332 Polina Makievsky. 2018. Rethinking six management mantras for better innovation. Stanford Social Innovation Review. https://ssir.org/articles/entry/rethinking_six_management_mantras_for_better_innovation.

333 Ernst & Young. No date. Connecting innovation to profit: Five key insights from the world's leading entrepreneurs.

334 Eleanora Ferrero. 2014. Where do innovative ideas come from? https://www.linkedin.com/pulse/20140803215938-33705437-where-do-innovative-ideas-come-from/.

porous boundaries that allow ideas to arise from anywhere.[335] And keep your key stakeholders in mind.[336]

Do not forget to provide resources to enable your organization's innovation.[337,338] Allocate flexible funding. Structure management and non-management staff roles to enable innovation. And invest in training, tools, and technologies.

The Innovation Process

Turning an idea into an innovation is not a trivial matter. It is a long and complex journey toward a seemingly unattainable goal. Structure simplifies the journey, providing understandable, achievable waypoints propelling you to your destination.

The innovation process often has multiple structures ranging from three to seven steps.[339,340,341,342] The following process has six steps informed by others' views but unlike any individual example.

As you review the process, keep three fundamentals in mind. First, share the process with all staff and encourage them to use it as a guide for generating ideas. Second, do not be deterred that your organization is too small to

335 Nidhi Sahni, Laura Lanzerotti, Amira Bliss, and Daniel Pike. 2017. Is your nonprofit built for sustained innovation? August. The Bridgespan Group. https://www.bridgespan.org/insights/library/strategy-development/innovation-capacity-summary-and-resources.

336 Polina Makievsky. 2018. Rethinking six management mantras for better innovation. Stanford Social Innovation Review. https://ssir.org/articles/entry/rethinking_six_management_mantras_for_better_innovation.

337 Daniel Cohen, Brian Quinn, and Erik Roth. 2019. The innovation commitment. October. McKinsey Quarterly. https://www.mckinsey.com/business-functions/strategy-and-corporate-finance/our-insights/the-innovation-commitment.

338 Nidhi Sahni, Laura Lanzerotti, Amira Bliss, and Daniel Pike. 2017. Is your nonprofit built for sustained innovation? August. The Bridgespan Group. https://www.bridgespan.org/insights/library/strategy-development/innovation-capacity-summary-and-resources.

339 Gerald Haman. 2000. Fostering creativity and innovation as a process: How to implement ideas that produce profitable solutions. Journal of Innovative Management. Summer, pp. 31–38.

340 Alissa Mariello. 2007. The five stages of successful innovation. Sloan Management Review. 1 April. https://sloanreview.mit.edu/article/the-five-stages-of-successful-innovation/.

341 Lauren Landry. 2017. The innovation process: A step-by-step guide. Northeastern University Blog. 21 December. https://www.northeastern.edu/graduate/blog/innovation-process/.

342 Langdon Morris. 2013. How to innovate: The innovation process. Innovation Management. 8 August. https://innovationmanagement.se/2013/08/08/how-to-innovate-the-innovation-process/.

innovate. The process is relevant for organizations of all sizes. And third, be sure to integrate your innovation process with other organizational processes.

Strategic Alignment

The goal of innovation is to develop ideas that support your organization's efforts to achieve its vision, mission, and strategy. Those three elements bound your needs investigation and idea generation. Review your vision, mission, and strategy. If you do not have all three elements, postpone your innovation process until you do.

Investigate Needs

What opportunities exist to achieve your strategic goals? What issues need to be solved to do so? What unmet needs and frustrations are being ignored? Focus on the opportunities and problems and not the solutions.

If you were the end user what would make your experience better? Adopt the perspective of your product or service end-user. What might they need to improve their experience with your organization and their lives? Observe. Survey. Verify your thoughts and ask them for theirs. Employ your active listening skills.

Generate Ideas

Now is your chance to be creative. Let your strategy determine your scope and the results of your investigation inform you. This is the time for divergent thinking. Generate creative ideas by exploring many possible solutions. Do not criticize the ideas yet. Employ your group dynamics diagnosis skills to maintain group cohesion.

Your group will generate more ideas if you gather a group of diverse thinkers. The Whole Brain Model represents four thinking styles Ned Herrmann identified when he was tasked with making General Electric managers more creative.[343] The model maintains that each of us is thought to have a preferred thinking style.

343 Ned Herrmann and Ann Herrmann-Nehdi. 2015. The Whole Brain Business Book: Unlocking the Power of Whole Brain Thinking in Organizations, Teams, and Individuals. Second Edition. McGraw Hill. New York, New York.

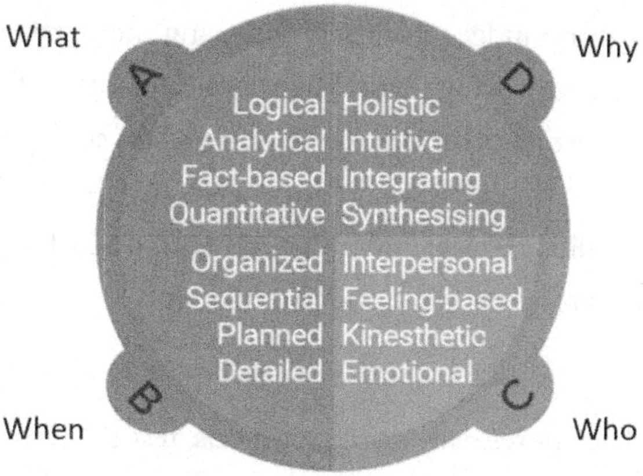

Source: Adapted from Ned Herrmann and Ann Herrmann-Nehdi. 2015.

1. The *A* quadrant thinker is logical, analytical, fact-based, quantitative, and interested in *what.*
2. The *B* quadrant thinker is organized, sequential, planned, detailed, and interested in *when.*
3. The *C* thinker is interpersonal, feeling-based, kinesthetic, emotional, and interested in *who.*
4. The *D* thinker is holistic, intuitive, integrating, synthesizing, and interested in *why.*

The future-based *D* thinkers tend to the most creative but the group benefits from others generating ideas that emphasize what, when, and who.

We can all use each quadrant, but we have preferences and preferences can change depending on the situation. A group representing all four thinking styles is more likely to generate a greater range of ideas than a group with similar thinking styles.

Evaluate Your Ideas

To evaluate your ideas, you switch from divergent thinking to convergent thinking. The group judges the ideas, criticizing them, refining them, combining them, and improving them. Many ideas are not worth implementing.

Identify the best ideas for this moment based on strategy, market, and financial and human resources. Evaluate benefits and risks to determine viability and sustainability. Be clear and transparent about the basis for eliminating ideas to preserve group cohesion and productivity.

Experiment

Once you have potentially viable solutions, test them. Construct a product, service, process prototype, or pilot and test it with staff and with loyal end users.

Testing and refinement allow your organization to learn about implementation and end-user receptivity. Allow the group enough time to analyze the results and modify the product, service, process prototype, or pilot. Repeat testing.

Implement

There are two goals for implementation. First, you must persuade your employees, all of them, to support the innovation. Help your employees understand what problem you were trying to solve, or opportunity you were trying to realize.

Show how the innovation is aligned with the organization's strategy and how the organization and employees will benefit. And help employees understand their role in implementing the innovation and your receptivity to suggested improvements.

Second, persuade your end users to accept the innovation. This requires changing end-user mental models so that they recognize the benefits.[344] Identify and clarify the underlying shift in thinking you need to accomplish. What

344 Mark Bonchek. 2014. Don't sell a product, sell a whole new way of thinking. 14 July. Harvard Business Review. https://hbr.org/2014/07/dont-sell-a-product-sell-a-whole-new-way-of-thinking.

was your original insight leading to the innovation? Why don't people understand your innovation? What is that moment when people change from disinterested to enthusiastic? Frame your persuasion as *from* to *better for the situation* rather than *bad* to *good*.

Determine how mental models are getting in the way of end users accepting your innovation. Do people fail to see the problem your innovation solves? Do people recognize the problem but fail to see how your innovation is a solution? Do people recognize the problem and the value of your solution but cannot make the change. Provide a roadmap that shows how to get from where they are now to where they want to go.

Be patient. Mental models do not change quickly. As people see how their new way of thinking benefits them, they will make the change.

PART 6
BUILD COMMUNITIES OF PRACTICE

PEERS, PARTNERS, AND HOSTING COMMUNITIES

We need others to complement and develop our own expertise.

Etienne Wenger, educational theorist and practitioner

What is a Community of Practice?

A community of practice is comprised of people with a shared interest who regularly come together to solve problems, improve skills, and learn from each other. Communities of practice are collaborative, supported by expertise, focused on a shared goal, and sustained over time.[345]

Members share best practices and explore new ways of doing things. Rooted in constructivism, the emphasis is on learning through complex problem analysis.[346] The collective knowledge of the group advances while advancing the individual member's knowledge.[347] Facilitation enables interaction and ensures safety and trust.

345 Etienne and Beverly Wenger-Trayner. 2015. Communities of practice: A brief introduction. 15 April. https://wenger-trayner.com/introduction-to-communities-of-practice/.

346 Saul McLeod. (2019, July 17). Constructivism as a theory for teaching and learning. Simply Psychology. https://www.simplypsychology.org/constructivism.html.

347 Christopher Johnson. 2001. A survey of current research on online communities of practice. Internet and Higher Education. 4:45–60.

Benefits

Communities of practice offer significant benefits.[348,349] A community of practice connects people across organizational, geographical, socioeconomic, and socioecological barriers providing a shared context for communication. Members are enabled to take collective responsibility for their own learning, be stimulated by others, and harvest a collective intelligence. And a community of practice creates a direct link between learning and performance helping people improve.

As a leader, you can facilitate the birth, growth, and sustainability of a community of practice. The members will ultimately define and sustain the community.

Five Stages in the Life of a Community of Practice

There are five stages in the life of a community of practice.[350,351] The first stage *defines the community*. Potential community members are identified and connect. Purpose and goals are defined.

In the second stage members *design the community* of practice. They learn about each other by sharing knowledge and experiences, develop a common vocabulary, and establish roles and processes.

Members *engage the community* in the third stage. They develop trust, commitment, and loyalty, model knowledge-sharing behavior, manage workflow, expand membership, test assumptions, and refine strategy.

The fourth stage *activates the community*. Members focus on solving problems, support and measure the community's work, and begin to rely on each other.

348 Etienne and Beverly Wenger-Trayner. 2015. Communities of practice: A brief introduction. 15 April. https://wenger-trayner.com/introduction-to-communities-of-practice/.

349 Darren Cambridge and Vicki Suter. 2005. Guide for designing and cultivating communities of practice in higher education. https://library.educause.edu/resources/2005/1/community-of-practice-design-guide-a-stepbystep-guide-for-designing-cultivating-communities-of-practice-in-higher-education.

350 Darren Cambridge and Vicki Suter. 2005. Guide for designing and cultivating communities of practice in higher education. https://library.educause.edu/resources/2005/1/community-of-practice-design-guide-a-stepbystep-guide-for-designing-cultivating-communities-of-practice-in-higher-education.

351 P. Gongla and C. Rizzuto. 2001. Evolving communities of practice: IBM global services experience. IBM Systems Journal 40(4):842–862.

The fifth and final stage in the life of a community of practice is *adapting the community*. Members work together to create new solutions, advance knowledge, and develop new capabilities.

Communities of practice can disband after solving a problem or persist indefinitely by addressing additional problems.

Successful Communities

Successful communities clearly identify their shared purpose, include a skilled facilitator, someone dedicated to nurturing the community, and offer clear and relevant value to their members.[352] Community members interact regularly and have a sense of ownership and a level of trust.

The community has an energized core group, shares or designates someone to be responsible for activities, and is clear about decision-making and accountability. Experts are part of the community or can be accessed when needed. Adequate resources negate barriers to participation.

Types of Communities of Practice

Communities of practice can take many forms. Three forms with representative structures and levels of investment are discussed in this chapter – Peer-to-Peer Consults, Partner Meetings, and Hosting Communities. The following two chapters extend the concept to citizen science and socioecological systems.

Peer-to-Peer Consults

A peer is equal to you in abilities, qualifications, age, background, or social status. A peer-to-peer consult is a natural, safe, and focused mechanism for you or your organization to connect with other practitioners, learn from their experiences, and develop new ideas.

Peer-to-peer consults can be applied within a single organization or among independent professionals who share your practice area. The process

352 Etienne Wenger, Richard McDermott, and William Snyder. 2002. Cultivating Communities of Practice. Harvard Business School Press. Brighton, Massachusetts.

described below is an adaptation of a 2013 Conservation Leadership Programme exercise led by Maureen Ryan, Christine Ageton, and Robyn Dalzen, inspired by Cambridge Leadership Associates' intervention model.[353]

Basic Peer-to-Peer Consult

Four to five peer consultants guide a conversation that helps a case presenter address a problem. One of the consultants facilitates, one keeps track of time, and another takes notes. The consultation takes one hour and emphasizes gathering information and analyzing the problem. The process has 6 steps.

Step 1 Problem Presentation – 5 Minutes

The case presenter tells their peers about the problem they are facing. The brevity of the presentation forces the presenter to focus on the most important information.

The case presenter should at a minimum objectively identify the stakeholders and their interests, and describe the problem and what actions they have taken or plan on taking. They should conclude by presenting the question they want the peer consultants to address.

Step 2 Problem Clarification – 10 Minutes

The peer consultants have 10 minutes to ask questions to clarify the problem. They should focus on critical questions as this is the only opportunity to query the case presenter. The peer consultants should not begin to analyze the problem.

Some of the questions might challenge the presenter. This is intentional, as the case presenter gains more when the consulting team is authentic and respectfully critical. Peer consultant questions clarify issues raised in the presentation and raise issues not presented. For example,

353 Ronald Heifetz, Alexander Grashow and Marty Linsky. 2009. The Practice of Adaptive Leadership. Harvard Business Press. Cambridge, Massachusetts.

- Is any critical information missing?
- What is at stake?
- What does success look like for you or your organization?
- Have you or your organization contributed to the problem?
- Who are the stakeholders?
- What is their perspective?
- Do any of the stakeholders have authority?
- Are there relationships between stakeholders?
- What does success look like to the stakeholders?

Step 3 Problem Analysis – 15 Minutes

The presenter goes to the *balcony* by pushing his/her chair outside the group. The balcony is a concept of Cambridge Leadership Associates and affords the presenter a distanced perspective. From the balcony the presenter remains quiet and listens while the peer consultants analyze the problem.

The peers discuss the problem openly and honestly.

- What seems to be at the root of the problem?
- What information is the presenter not considering that might be helpful?
- What has the presenter tried and what was the result?
- What are the barriers to other actions?
- Are some options off the table?

Step 4 Possible Actions – 10 Minutes

The presenter remains on the balcony while the peer consultants discuss possible actions to resolve the problem.

- What possible actions can the presenter take?
- What conversations with stakeholders need to happen?
- What new partnerships or relationship shifts need to occur?
- Are there any low-risk experiments that could be tried?
- What specific goals could be achieved soon?

Step 5 Presenter Reflection – 10 Minutes

The presenter returns to the group and shares their experience on the consult. What was it like to have others assess your problem? Did you learn anything about yourself? Are any peer-recommended actions actionable? Did the consultation inform your problem-solving?

Step 6 Presenter and Peer Consultants Debrief – 10 Minutes

The presenter and peer consultants reflect on how they contributed to the problem-solving process. In the spirit of appreciative inquiry, everyone should name two aspects they did well and one aspect they could improve. For example,

- Did the presenter provide the needed information in an unbiased fashion?
- Was the presenter receptive to the analysis and recommendations?
- Were the peers respectfully candid, focused on helping, objective, and circumspect?

Finally, comment on the utility of the peer-to-peer consult and how it might be improved.

Enhancing Your Peer-to-Peer Consult

In the chapter on communication, you learned about nonverbal communication. In the peer consultant role, the discussion about voice intonation, pace, and amplitude, and gestures can help you understand the presenter's meaning. Other clues will help you better understand the presenter's problem and the presenter's level of objectivity.

- Does the presenter use language that exaggerates or understates the problem?
- Is the presenter trying to make everyone happy rather than solve the problem?

- Is the presenter speaking dispassionately or with emotion?
- How does the presenter portray other stakeholders?

The presenter framed their problem for the peer consultants. The framing helps the presenter simplify and make meaning out of the situation but that framing can introduce subjectivity and result in adherence to a story that inhibits problem understanding.

Common stories include being a hero fighting the problem, us versus them, believing they are the only one who knows the truth. A presenter might fashion a problem so large it is unsolvable. Or the presenter might choose to be the objective mediator rather than a stakeholder or be missing from the story. All of these are clues to a presenter's blind spots.

Partner Meetings

A partner is someone who shares or is associated with you and your organization in some action or endeavor. Partners share information or funds, collaborate, and serve as influential supporters and important advisors. The relationship is mutually beneficial and enduring operating most often at the organization's strategic level.

Partners may or may not be entirely aligned in their goals but have committed to an honest, productive relationship that serves the interests of all parties to the extent possible. Partner meetings Create or enhance partnerships, seek input on program direction, promote coordination among partners, and encourage partner investment.

Partner meetings are a process, not a single event. Their structure and process reflect the goals of the partners. Two anonymous exemplars illustrate the benefit of partner meetings.

Exemplar One – Sector Diversity

A state agency is the source of reliable, objective information and expertise to support stewardship of native species and habitats. The organization serves public agencies, the private and education sectors, and communities.

The organization's 15 founding partners included state and federal agencies, a forest products company, the local chapters of two global environmental nonprofits, a land trust, and a structural engineering firm. Partners are collaborators, beneficiaries of the organization's information, or both. The meetings occur once a year and initially focused on the organization's activities. The number of partners has increased to 20 organizations.

Over time the partner meeting evolved to include opportunities for the partners to discuss their own organization's goals and problems. The more balanced agenda enabled the organization to co-develop strategy that met its mission and better served its partners. In essence, the partners have become a network with all members benefitting.

The partner meetings have produced significant benefits for the agency. Tangible benefits include collaborative pursuit of increased core funding from the legislature and increased core and project funding from the partners.

Partner meetings are improved by actions signaling the agency's respect and caring for the other partners, made evident by pre- and post-meeting meetings. The program coordinator meets individually with partners in advance of the annual partner meeting to obtain feedback about the agenda and to seek advice about sensitive, significant, and complex topics. This ensures the partner feels valued, attends the partner meeting prepared, and avoids unpleasant surprises for both parties.

The program coordinator also meets individually with the partners after the meeting to reinforce collective commitment to, and ownership of, the process and goals. The partners also follow-up on specific items

Exemplar Two – Shared Purpose

The second organization is a university-sponsored nonprofit that advances the conservation of the state's native species and ecosystems through science, planning, and education. They are the only comprehensive source of information on the status and location of the state's rarest and most threatened species and natural plant communities. They share the information with a wide range of stakeholders in partnerships that work to ensure biodiversity resources are not unknowingly diminished.

In contrast to the first exemplar's partner meeting purpose, the primary purpose of this initial partner meeting was to connect key members of the state's conservation community. Secondary goals were to provide a forum for communication, enhance collective effectiveness, highlight successes and important developments, and determine which priorities would benefit from collective action.

Thirty-two organizations were represented at the initial meeting. The agenda included:

- Updates from the partners.
- Highlights of successful partnerships.
- Roundtable discussions about specific conservation topics.
- A wrap-up session about how everyone could do a better job achieving conservation goals.

Attendee response was overwhelmingly positive producing a renewed sense of purpose and a desire to maintain momentum. One hundred percent of the attendees said they would like to meet again, and almost without exception, once a year. Partners supported moving the location among different venues, and identified a suite of issues they would like to address.

Both exemplar leaders believe that partner meetings are extremely valuable, economical, and an effective tool for organizational success. The meetings facilitate the development of a robust network of partners and supporters and create opportunities for network growth. In their opinion, the investment in planning, follow up, and periodic check-ins during the year pays for itself.

Hosting Communities

Sustaining healthy lands and waters is a complex problem with no simple answers. No one individual can find and implement a solution. Hosting draws on the collective intelligence of communities while creating advocates and identifying people willing to help with implementation.

Hosting is a practice for harnessing the collective wisdom and self-organizing capacity of groups of any size.[354] Leaders as hosts promote meaningful conversations among people to gather new insights for action.[355] As a host you are neither a moderator nor a conventional facilitator. Instead, you guide the deep and meaningful conversations of others, and harvest their collective intelligence.

The benefits of hosting include better decision-making, capacity-building, a more rapid response to problems and change, and participants empowered and committed to implementation.

Hosting is used to analyze issues, resolve conflicts, and develop strategy. Hosting requires:

- Organizational, communication, and analytical skills, supported by empathy and the ability to shape seemingly disparate thoughts into a single, understandable idea.
- The ability to frame strategic questions that encourage possibilities rather than problem-solving.[356]
- Facilitating working conversations that enhance trust and reduce fear.
- Connecting diverse perspectives.
- Asking what is possible rather than focusing on what is not working.
- Fostering shared meaning by highlighting commonalities.
- Nurturing knowledge acquired through practice.

Hosting begins with a clear purpose, selecting and supporting a hosting team, identifying a powerful question, designing a compelling invitation, and

354 Art of Hosting. No date. Art of hosting and harvesting conversations that matter. Artofhosting.org.

355 Deborah Frieze and Margaret Wheatley. 2010. From hero to host: A story of citizenship in Columbus, Ohio. https://berkana.org/berkana_articles/from-hero-to-host-a-story-of-citizenship-in-columbus-ohio/.

356 Eric Vogt, Juanita Brown, and David Isaacs. 2003. The art of powerful questions: Catalyzing insight, innovation, and action. Whole Systems Associates. Mill Valley, California.

preparing the meeting space and the participants.[357] Participants can include anyone with a stake in the issue.

The hosting team decides which hosting method is most likely to create a comfortable environment and produce meaningful conversation. Two methods are described below. During the meeting, the team guides the process and supports the participants in its quest to collectively learn and discover, then harvesting what the group produces.

Circle

Circle is a method unto itself and a component of other methods. As a method, circle is effective for creating collaborative dialogue for groups of 5 to 25. The circle shape enables eye-to-eye contact, signals equality, and creates a dynamic learning opportunity.

Circle fosters a sense of partnership where everyone knows they are heard. Circle gives people a way to influence a conversation, empowers the participants, and enhances group cohesiveness and effectiveness. Circle leadership is shared and lightly held. Roles include:

Host

A host establishes the circle's intention by developing a powerful question, defining expected outcomes, determining participants, setting an agenda, issuing the invitations, preparing the space, and helping define the scope of the conversation. For multiple meetings, the host position rotates among participants as they become familiar with the process.

Guardian

A guardian works with the host to keep the circle centered and intentional. Notably, the guardian aids self-governance and maintaining group intention by observing process. They call a pause in the conversation using a gentle

357 Art of Hosting. No date. Art of hosting and harvesting conversations that matter. Artofhosting.org.

noisemaker or other device to refocus, note a shift, or capture a decision. Any participant can request a pause.

Scribe

A scribe, typically a circle participant, records insights, decisions, and actions. The scribe can be a single individual or rotated throughout the circle. The recording can be written notes or graphics and is made available to participants after circle.

Participants

Circle participants accept shared responsibility for the quality of the experience and agree to focus on group cohesion and productivity rather than personal agendas. Participants speak with intention and relevance, listen with attention, and tend to the well-being of the group.

A circle is executed in six steps.[358] The *start point or welcome,* is the responsibility of the host or guardian. When participants gather, they are brought to attention with a gesture such as a moment of silence, meditation, a reading, or music.

Setting the center helps people remember they are part of a group. An object that represents intention is placed in the center of the circle.

Check-in or greeting, helps people adopt a frame of mind for conversation and reminds everybody of their commitment to the meeting's intention. This is typically accomplished by verbal sharing starting with a voluntary participant and proceeding around the circle.

Setting circle agreements enables the trust and courage to have a productive conversation. Agreements include respect, compassion, and curiosity for other views, responsibility for the well-being of the group, confidentiality, and asking for what we need and offering what we can.

358 The Circle Way. Accessed 4 February 2010. The circle way: Basic guidelines for calling a circle. http://www.thecircleway.net/.

Holding the conversation centers on the powerful question posed by the host. The guardian keeps the circle centered and intentional and the scribe records insights, decisions, and actions.

Check-out and farewell allow a few minutes for each participant to reflect on what they learned. The host or guardian may offer a few words in summary or thanks before the circle is released.

World Café

World café is an expansion of circle able to accommodate hundreds of people. It is well suited to determining what a community is thinking and feeling.[359]

Groups of 4 to 6 participants gather around tables to discuss a matter of importance. After each conversation, a host remains at the table while the others disperse to new tables. This enables the threads of the various conversations to be woven together. Participants gain a sense of the collective discoveries.

World café has seven design principles.[360]

1. Set the Context by Establishing the Purpose and Parameters. What questions will be asked? Who will participate? What is proper participant etiquette? How will the results of the conversations be harvested?
2. Create Hospitable Space That Feels Safe and Inviting. How can you make participants comfortable to think, speak, and listen? How will your invitation and physical setting contribute?
3. Explore Questions that Matter so that Knowledge Emerges in Response to Your Questions. Will you have one question or multiple questions? What questions are relevant to the concerns of the

359 Art of Hosting. Accessed 4 February 2020. Art of hosting and harvesting conversations that matter. Artofhosting.org.

360 The World Café. 2015. Café to go! A quick reference guide for hosting world café. The World Café Community Foundation. http://www.theworldcafe.com/wp-content/uploads/2015/07/Cafe-To-Go-Revised.pdf.

participants? Will your questions spread from the café to the community at large?

4. Encourage Everyone's Contribution to Optimize Learning and Collective Intelligence. How can you create an environment that promotes participants adding to the conversation rather than just listening?
5. Connect Diverse Perspectives to Increase the Possibility of New Insights. How do you ensure diversity among participants? How can you encourage participants to carry their ideas from table to table? How can you encourage participants to listen with respect and without bias?
6. Listen Together for Patterns and Insights to Understand Central Ideas. How do you encourage listening for meaning? How can you help participants identify the key themes, patterns, and insights arising from the conversations?
7. Share Collective Discoveries to Harvest the Collective Intelligence of the Participants and Make it Available to Everyone. How can you summarize the key themes, patterns, and insights arising from the conversations? How can the participants take the themes, patterns, and insights with them to continue reflecting and to share with others?

To host a world café seat four or five people at small, round tables. Each table should have markers and index cards or be covered with paper. Plan on at least three rounds of conversation around a powerful question. Encourage participants to speak, write, or illustrate their ideas on a card or the paper table covering. Each round should last about 20 minutes.

On completion of a round, the table host remains to provide group-to-group continuity and facilitate identification of key themes, patterns, and insights. The other participants distribute themselves among other tables to discuss the same question with new participants.

After several rounds this cross-pollination should begin to connect and make explicit ideas, patterns, and themes. Participants return to their initial table for the last round to synthesize their learning and thinking.

Harvest the learning of the participants in a full group conversation. Provide a few minutes for self-reflection on key ideas, patterns, themes, and any deeper questions that arose. Ask participants individually or via a table host to share with the larger group. All contributions should be recorded and made available to participants. The world café can be extended or another café scheduled to answer deeper questions.

Making collective knowledge visible is an essential element of world café.[361] The paper tablecloths can be attached to the wall enabling the participants to take a gallery tour. Participants can create affinity clusters using post-it notes to help with next steps. You can enlist a graphic recorder who draws the group's ideas on flip charts or a wall mural using text and graphics. Or the host can create a storybook using graphics and text.

Craft Compelling Questions for Circle and World Cafe

Compelling questions attract energy, focus attention, invite inquiry, and surface new ideas.[362,363,364] They are simple, clear, thought-provoking, and open new possibilities. They invite people to communicate new ideas.

Compelling questions speak personally, share stories, express hope, address uncertainties and complexities, and expand people's range of experience. You can create compelling questions by focusing on three dimensions, each helping people change their thinking and inspiring a sense of possibility that encourages action.[365]

361 The World Café. 2015. Café to go! A quick reference guide for hosting world café. The World Café Community Foundation. http://www.theworldcafe.com/wp-content/uploads/2015/07/Cafe-To-Go-Revised.pdf.

362 The World Café. 2015.

363 Eric Vogt, Juanita Brown, and David Isaacs. 2003. The art of powerful questions: Catalyzing insight, innovation, and action. Whole Systems Associates. Mill Valley, California.

364 Maggie Herzig and Laura Chasin. 2018. Fostering dialogue across divides: A nuts and bolts guide from Essential Partners. https://whatisessential.org/resources.

365 Eric Vogt, Juanita Brown, and David Isaacs. 2003. The art of powerful questions: Catalyzing insight, innovation, and action. Whole Systems Associates. Mill Valley, California.

Why Questions

The first dimension of a compelling question is construct. *Why* questions, and to a lesser extent *how* questions, stimulate more reflective and creative thinking than *which, who, when*, and *what* questions. Why questions should reflect genuine curiosity and not be accusatory.

Scope

The second dimension of compelling questions is scope. A compelling question must match its intent and people's capacity to act. People will not engage if the question is irrelevant or beyond their control.

Unbiased

Finally, compelling questions should not make assumptions or include unconscious beliefs that unintentionally constrain ideas or cause participants to disengage. For example, do not assume everyone believes that healthy lands and waters can only be achieved by shutting down the fossil fuel industry. Test your assumptions and beliefs before including them in a question and reframe your question to encourage exploration.

ENGAGE CITIZEN SCIENTISTS

Science and everyday life cannot and should not be separated.

Rosalind Franklin, English chemist and X-ray crystallographer whose work was central to the understanding of the molecular structures of DNA, RNA, viruses, coal, and graphite

Citizen Science

Citizen science is public participation and collaboration in scientific research for the purpose of increasing knowledge. Anyone with capacity, curiosity, or concern can be a citizen scientist, helping professional scientists identify research questions, collect and analyze data, interpret results, and develop solutions.

Ideally, citizens and scientists work together using a shared protocol that produces reliable data that can be combined with other data. The data and results are then shared with all.

Citizen science originated at least two millennia ago when the Chinese public helped track migratory locust outbreaks.[366] The role of the public in biological science diminished in the 18th century with the development of Linnaean binomial nomenclature and concomitant rise of professional training.

366 Tian Huidong et al. 2011. Reconstruction of a 1,910-y-long locust series reveals consistent associations with climate fluctuations in China. Proceedings of the National Academy of Science 108(35):14521–14526.

Citizen science bird count projects and other ecological studies beginning in the early 20th century signaled a citizen science revival.[367,368,369]

Alan Irwin, with interests in science and technology policy and environmental sociology, coined the phrase *citizen science* in the 1990s.[370] He defined citizen science as *science which assists the needs and concerns of citizens ... a form of science developed and enacted by the citizens themselves*. Present day citizen science supports basic and applied ecology including species range shifts, phenology, community composition, habitat loss, invasive species, infectious disease, and biocontaminants.[371]

Modern citizen science is breaking boundaries through the sheer volume of data and participants. The world's largest repository of biodiversity data, the Global Biodiversity Information Facility, reports that 800 million of its 1.6 billion data points come from nonprofessionals.[372]

Muki Haklay, Co-director of the University College London Extreme Citizen Science Group, estimated 72,000 directly and significantly engaged citizen scientists in the United Kingdom.[373] Assuming citizen scientists are connected to the Internet and have tertiary educations, Haklay estimates 1.7 million citizen scientists worldwide.

The assumptions about connectedness and education seem to be at odds with citizen science efforts in developing economies. Extrapolating directly from the United Kingdom ratios and a global population of 7.7 billion, the upper boundary for current citizen scientists is likely about 8.5 million. The

367 Abraham Miller-Rushing, Richard Primack, and Rick Bonney. 2012. The history of public participation in ecological research 10(6):285–290.

368 Jane Gregory and Steve Miller. 1998. Science in Public: Communication, Culture, and Credibility. Basic Books, Perseus Books Group. Cambridge, Massachusetts.

369 Jack Stilgoe, Simon Lock, and James Wilsdon: 2014. Why should we promote public engagement with science? Public Understanding of Science 23(1):4–15.

370 Alan Irwin. 1995. Citizen Science: A Study of People, Expertise, and Sustainable Development. Routledge, Taylor and Francis Group. London.

371 Janis Dickinson, Benjamin Zuckerberg, and David Bonter. 2010. Citizen science as an ecological research tool: Challenges and Benefits. 11 April. Annual Review of Ecology, Evolution, and Systematics. 41:149–172.

372 Aisling Irwin. 2018. Citizen science comes of age. 25 October. Nature. 562:480–482.

373 Muki Haklay. 2018. How many citizen scientists in the world? Personal blog. 5 October. https://povesham.wordpress.com/2018/10/05/how-many-citizen-scientists-in-the-world/.

number of potential citizen scientists is much greater, and combined with more than 7 million researchers, more than adequate for keeping our lands and waters healthy.

Benefits of Citizen Science

The use of citizen science depends on its acceptance by the scientific community. Recent research on methodology and validation techniques supports increased reliance on citizen science, particularly at large scale.[374,375] And advances in techniques, communications, technology, transportation, and computing facilitate scientific and educational outcomes.[376]

Citizen science offers leaders three types of benefits. First, citizen scientists extend resources for filling data gaps. Project scopes normally beyond your reach are now attainable. For example, the Audubon Christmas Bird Count relies on volunteers, more than 80,000 in 2020.

And because citizen scientists are typically volunteering the cost-benefit ratio is extremely favorable. Expenses are limited to materials, equipment, and any additional perks you choose to provide, expenses that can be included in your budget submitted to the funder.

Second, citizen science facilitates learning for all parties. The hands-on nature of citizen science helps the brain understand meaning in relation to other meanings it experienced, rather than as new, discrete entries.[377]

Citizen science promotes collective learning and overcomes our mental model bias to reject information at odds with current experience.[378] And citizen science is aligned with the cognitivism and constructivism that form the

374 Ria Follett and Vladiir Strezov. 2015. An analysis of citizen science-based research: Usage and publication patterns. PLoS ONE 10(11): e0143687.

375 Janis Dickinson, Benjamin Zuckerberg, and David Bonter. 2010. Citizen science as an ecological research tool: Challenges and Benefits. 11 April. Annual Review of Ecology, Evolution, and Systematics. 41:149–172.

376 Abraham Miller-Rushing, Richard Primack, and Rick Bonney. 2012. The history of public participation in ecological research 10(6):285–290.

377 Ben Thomas. 2012. Meaning on the brain: How your brain organizes reality. https://blogs.scientificamerican.com/guest-blog/meaning-on-the-brain-how-your-mind-organizes-reality/. 26 December.

378 Dedre Gentner and Albert Stevens (Eds.). 1983. Mental Models. Lawrence Erlbaum Associates. Mahwah, New Jersey.

basis for next generation science standards and the universal design for learning.[379,380]

The third major benefit of citizen science is the development of socially meaningful relationships that facilitate the implementation of knowledge gained through collaboration. Citizen scientists develop a sense of ownership and accomplishment. They are motivated to advance research, contribute to solutions, and to influence larger decisions about science policy.

Challenges of Citizen Science

Citizen science is not yet fully accepted among scientists and policy makers as a reliable producer of unbiased, reliable information. Acceptance will continue to increase as project models are refined and adopted and success stories are communicated.

Individual and community participation in citizen science is uneven. Relevancy and trust are critical. People will not invest significant effort unless the question being addressed is clear, relevant, and a priority. Cooperating scientists must be accepted as experts with the individual's or community's interests at heart.

Data quality is the most quantifiable challenge for citizen science. Valid data is an important scientific outcome. Citizen scientists also want to know that their work is meaningful and that they made a valuable contribution toward answering the research question.

Factors affecting data quality include citizen scientist variability in age, experience, skill, training, and willingness to adhere to protocols.[381] Many of the same factors that affect data quality of professional projects. Advancements

379 National Science Teaching Association. No date. About the next generation science standards. Accessed 20 February 2020. https://ngss.nsta.org/about.aspx.

380 Amanda Morin. 2019. Universal design for learning: What you need to know. Understood. https://www.understood.org/en/learning-thinking-differences/treatments-approaches/educational-strategies/universal-design-for-learning-what-it-is-and-how-it-works.

381 Caren Cooper, Wesley Hochachka, and Andre Dhont. 2012. The opportunities and challenges of citizen science as a tool for ecological research, pp. 99–113, In Citizen Science: Public Participation in Environmental Research (Eds. Janis Dickenson and Rick Bonney). Comstock Publishing Associates, a Division of Cornell University Press. Ithaca, New York.

in citizen science protocols and data evaluation are reducing concerns about citizen science data quality.

Levels of Citizen Science Engagement

The disparate understanding of what citizen science is affects definitions of engagement. For example, Shirk et al. identify five engagement models.[382]

1. *Contractual projects* - Citizen scientists (e.g., communities) ask professionals to conduct a specific inquiry and report results.
2. *Contributory projects* - Citizen scientists contribute data to research designed by professionals.
3. *Collaborative projects* - Citizen scientists help professionals refine project design, contribute data, analyze data, and disseminate findings.
4. *Co-created projects* - Citizen scientists collaborate with professionals on project design, data collection, analysis, and interpretation, and disseminating findings.
5. *Collegial contributions* – Nonprofessionals conduct independent research that may or may not be used by professionals.

Haklay's perspective is more expansive and hopeful, including seven levels of science engagement.[383]

1. *Everyone* - Science impacts society so everyone is exposed.
2. *Passive consumption of science* - Through newspapers, websites, television, and radio.
3. *Active consumption of science* - Visits to science and natural history museums.
4. *Active, but limited, engagement in citizen science* - Participation in a single task once per year.

382 Jennifer Shirk et al. 2012. Public participation in scientific research: Framework for deliberate design. Ecology and Society. 17(2):29. http://www.ecologyandsociety/vol17/iss2/art29.

383 Muki Haklay. 2018. How many citizen scientists in the world? Personal blog. 5 October. https://povesham.wordpress.com/2018/10/05/how-many-citizen-scientists-in-the-world/.

5. *Remote engagement* - Online volunteer computing or thinking.
6. *Data collection and analysis* - Regular data collection.
7. *High engagement in do-it-yourself science* - Exploring do it yourself science and developing your own project.

Leaders should encourage both perspectives as any interest in science is better than no interest in science. Even minimal engagement - entry at a level that matches needs, interests, and abilities - creates an opportunity for deepening participation.

Participation escalation - the increasing engagement of individuals or communities in citizen science - was a fundamental element of the *Doing It Together Science* project of the European Union.[384] The project goal was to enhance the awareness of citizen science and do it yourself science at the public and policy levels. Through 860 events over a 3 year period the project reached more than 500,000 people.

The citizen science models best able to meaningfully advance knowledge and learning include direct and significant collaboration between citizen scientists and professional scientists - contributory, data collection and analysis, collaborative, and co-created projects.

A Framework for Contributing, Collaborative, and Co-created Citizen Science Projects

Citizen science is a reasonable choice when 1) volunteers can collect high-quality data, 2) volunteers are the only way to address the research question, and 3) public participation serves sociological and learning goals.[385] Shirk et al. developed a framework for public participation

384 Muki Haklay. 2016. Introducing Doing It Together Science: An EU citizen science project. 29 April. https://uclexcites.blog/2016/04/29/introducing-doing-it-together-science-an-eu-citizen-science-project/.

385 Duncan McKinley et al. 2015. Investing in citizen science can improve natural resource management and environmental protection. Issues in Ecology, The Ecological Society of America. Report Number 19.

in science based on the W.K. Kellogg Foundation's logic model.[386,387] The citizen science framework is adapted here to recapture aspects of the logic model and align it with the goal of healthy lands and waters.

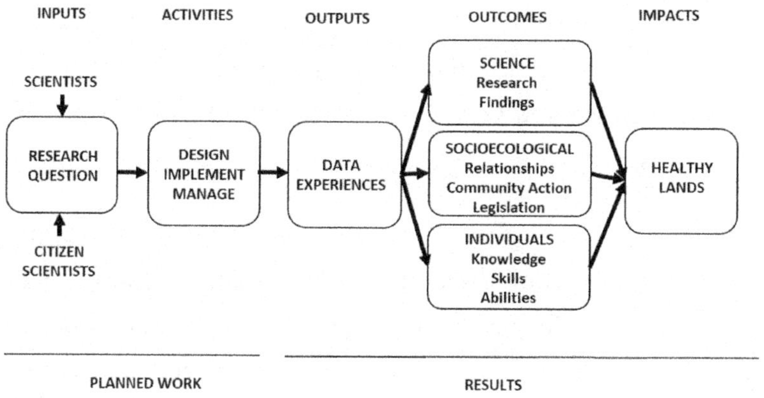

Source: Adapted from Jennifer Shirk et al. 2012.

Inputs

The first decision point is for the scientists to determine if the citizen science project is a contributory, collaborative, or co-created model. A contributory model project is potentially simpler to implement and can achieve science outcomes but is less likely to achieve significant socioecological and individual outcomes.

Collaborative and especially co-created projects potentially take longer to implement and manage. Their benefits include the potential to surface new information vital to developing the research question, motivating the citizen scientists, and attaining significant socioecological and individual outcomes in addition to the science outcome. Achieving healthy lands and waters is more likely with collaborative and co-created projects.

386 Jennifer Shirk et al. 2012. Public participation in scientific research: Framework for deliberate design. Ecology and Society. 17(2):29. http://www.ecologyandsociety/vol17/iss2/art29.

387 W.K. Kellogg Foundation. 2004. Logic Model Development Guide. W.K. Kellogg Foundation. Battle Creek, Michigan. https://www.wkkf.org/resource-directory/resource/2006/02/wk-kellogg-foundation-logic-model-development-guide.

Having decided which model to use, scientists in the case of contributory projects and scientists and citizen scientists together for collaborative and co-created projects, develop the research question. The research question is the *why* of the project and critical for success.

The interests of other parties such as other scientists, funders, and communities can be filtered and represented by the scientists and citizen scientists. The question should reflect the hopes, goals, and expectations of scientists and citizen scientists and be focused, understandable, and answerable using citizen scientists.

Activities

The research question drives the project design including where, when, how many citizen scientists for how long, transportation, food and water, safety, and funding. What training and oversight are required to ensure accurate data collection? Who manages the data? How will outputs be disseminated to facilitate outcomes and impact?

All these design questions are addressed by the scientists for contributory projects. Citizen scientists, except perhaps for funding, help to answer design questions for collaborative and co-created projects. Implementation and management follow project design.

Outputs

The activities of the project produce outputs. Typical outputs include observations, counts, measurements, and learning experiences. Clear, repeatable protocol, training, and management oversight determine the abilities and motivation of the citizen scientists. This in turn reflects on the quality of the outputs.

Outcomes

Outcomes are the changes you expect from your design, activities, and outputs. They are often short and intermediate changes and affect the citizen scientists. Outcomes involve interpretation of outputs and creating meaning through development of a coherent story. The story enables change.

In the case of citizen science for healthy lands and waters the changes could be in scientific knowledge or new knowledge, skills, and abilities imparted to the citizen scientists. Or the changes may be transformative activation of the sociological system in terms of increased awareness and stewardship, advocacy, policy, and legislation.

Impacts

Impacts are broader changes that occur within a community, organization, society, or environment because of project outcomes. They are long, sustainable changes that affect education systems, community and societal norms, and the health of the environment. For example, an impact might be that citizens understand the value of their community natural resources and develop education, policy, regulatory, and management practices to preserve that value.

WORK IN SOCIOECOLOGICAL SYSTEMS

Until society can be reclaimed by an undivided humanity that will use its collective wisdom, cultural achievements, technological innovations, scientific knowledge, and innate creativity for its own benefit and for that of the natural world, all ecological problems will have their roots in social problems.

Murray Bookchin, American pioneer in the ecology movement formulated and developed the theory of social ecology and urban planning

Socioecological Systems

Socioecological systems are linked systems of people and nature emphasizing that humans must be a part of, not apart from, nature.[388] Berkes and Folke developed a socioecological system framework to address questions about the resilience of each part of the system and how the two subsystems can be combined for mutual benefit. The concept has been applied to many disciplines including studies of the adaptability of socioecological systems to meet change without compromising long-term sustainability.[389]

388 Fikret Berkes and Carl Folke. 1998. Understanding dynamics of ecosystem-institution linkages for building resilience. Beijer Discussion Paper No. 112. The Beijer Institute of Ecological Economics, Royal Academy of Sciences. Stockholm, Sweden.

389 Fikret Berkes, Johan Colding, and Carl Folke (Eds). 2002. Navigating Social Ecological Systems: Building Resilience for Complexity and Change. Cambridge University Press. Cambridge, United Kingdom.

Sustaining healthy lands and waters is a complex challenge that cannot be addressed with top-down, command and control management that too often results in unforeseen consequences.[390] Success requires innovation, diverse knowledge, recognition of competing values, and building collective capacity.[391,392]

A socioecological system framework encompasses diverse values and perspectives through public engagement resulting in improved decision-making and equitable governance.[393,394,395] A socioecological system also promotes social learning - learning within and by interacting groups[396]. Learning among groups fosters the adaptive co-management necessary to address socioecological complexity.[397,398]

The collaborative nature of socioecological systems produces groups of individuals of different capabilities, interests, and intentions. A range of leadership qualities are essential for success.[399] Two qualities are particularly important.[400]

Boundary spanners connect the various individuals and disciplines building mutual trust and support through transformational changes in management and perceptions. Network weavers directly or indirectly create,

390 Crawford Holling and Gary Meffe. 1995. Command and control and the pathology of natural resource management. Conservation Biology 10(2):328–337.

391 Richard Norgaard 1989. The case for methodological pluralism. Ecological Economics. 1:37–57.

392 Richard Bawden. 1991. Systems thinking and practice in agriculture. Journal of Dairy Science. 74:2362–2373.

393 Daniel Yankelovich. 1991. Coming to Public Judgement: Making Democracy Work in a Complex World. Syracuse University Press. Syracuse, New York.

394 Mark Reed. 2008. Stakeholder participation for environmental management: A literature review. Biological Conservation. 141:2417–2431.

395 Jon Elster (Ed.). 1998. Deliberative Democracy. Cambridge University Press. Cambridge, United Kingdom.

396 Melanie Muro and Paul Jeffrey. 2008. A critical review of the theory and application of social learning in participatory natural resource management processes. Journal of Environmental Planning and Management. 51:325–344.

397 Per Olsson, Carl Folke, and Fikret Berkes. 2004. Adaptive co-management for building resilience in social ecological systems. Environmental Management. 34:75–90.

398 Emma Tompkins and W. Neil Adger. 2004. Does adaptive management of natural resources enhance resilience to climate change? Ecology and Society 9(2):10 https://www.ecologyandsociety.org/vol9/iss2/art10/.

399 Frances Westley et al. 2013. A theory of transformative agency in linked social ecological systems. Ecology and Society. 18(3):27. https://www.ecologyandsociety.org/vol18/iss3/art27/.

400 Orjan Bodin. 2017. Collaborative environmental governance: Achieving collective action in social ecological systems. 18 August. Science. 357(6352): eaan1114. https://science.sciencemag.org/content/357/6352/eaan1114.

intervene in, and shape the group as it evolves. Actions include brokering the group governance structure and creating collaborative venues.

Common Pool Resources

At the heart of socioecological systems are common pool resources. Common pool resources benefit a group of people but are diminished if everyone pursues his or her own self-interest. Examples of common pool resources include fisheries, forests, aquifers, and irrigation systems. Climate change highlights the common pool nature of our atmosphere.

Ecologist Garrett Hardin believed overuse and collapse of common pool resources were inevitable as population density increased.[401] Population density has indeed increased but fortunately Hardin's dystopian view has been moderated. Not just by government policies and agency action, but by hundreds of examples of people cooperating and developing rules that coordinate and limit use of common pool resources.[402,403,404]

Private and family forest landowners are managing forests for health, habitat, reduced fire risk, and wood products. US Regional Fisheries Management Councils are adopting an ecosystem approach to fishery quotas. Agriculture is using more efficient irrigation practices and tailoring crops to soil and moisture conditions.

Under what conditions will common pool resource users cooperate? Economist Elinor Ostrom determined that common pool resource users must address supply, commitment, and monitoring.[405]

401 Garrett Hardin. 1968. The tragedy of the commons. 13 December. Science. 162:1243–1248.

402 U.S. National Research Council. 1986. Proceedings of the Conference on Common Property Resource Management, April 21–26, 1985. Panel on Common Property Resource Management. National Academy Press. Washington, D.C.

403 Bonnie McCay and James Acheson (Eds). 1987. The Question of the Commons: The Culture and Ecology of Communal Resources. University of Arizona Press. Tucson, Arizona.

404 Fikreet Berkes (Ed.). 1989. Common Property Resources: Ecology and Community-based Sustainable Development. Belhaven Press. London, United Kingdom.

405 Elinor Ostrom. 1990. Governing the Commons: The Evolution of Institutions for Collective Action. Cambridge University Press. Cambridge, United Kingdom.

Successful common pool resource situations are recognized by attributes of the resource and the users.[406]

Favorable common pool resource attributes include:

- Resources are used enough to warrant organizing.
- Resource condition, if deteriorated, can feasibly be improved.
- Reliable and valid indicators of resource condition are available at a relatively low cost.
- The flow of resource units (e.g., fish, trees, water) is relatively predictable.
- The resource area is small enough that users can develop accurate knowledge of internal microenvironments and external boundaries.

Favorable common pool resource user attributes include:

- Users are dependent on the resources for a significant part of their livelihood.
- Users have a shared understanding of how the resource operates and how their actions affect each other and the resource.
- Users can predict the future value of the resource with relative certainty.
- Users trust each other and practice reciprocity.
- Users can establish access and harvesting rules without external authorities countermanding them.
- Users possess at least minimal organization and leadership skills.

These attributes are considered conditions positively related to the emergence of collective action. The more attributes present the greater the likelihood of collective action. The value of the attributes is likely to change in response to the institutional setting (e.g., government, academic, nonprofit).

406 Elinor Ostrom. 2000. The danger of self-evident truths. Political Science and Politics 33(1)33–44.

Favorable Common Pool Resource Design Principles

According to Ostrom, users are more likely to commit to collective action across generations if the arrangements are characterized by eight design principles.[407]

1. The boundaries of the resource, and the users with the right to withdraw from the resource, must be clearly defined.
2. Resource use rules defining withdrawal time, place, technology, and quantity are related to local conditions, and to rules requiring user commitments to managing the resource.
3. Individuals affected by the rules can participate in modifying them.
4. Monitors of common pool resource condition and user behavior are either the users, or accountable to the users.
5. Users who violate the rules can be sanctioned by other users, and officials accountable to the users.
6. Users and their officials have rapid access to low-cost, local mechanisms for resolving conflicts.
7. The rights of users to devise their own institutions are not challenged by external government authorities.
8. For common resource pool groups part of larger systems, use, management support, monitoring, enforcement, conflict resolution, and governance activities are organized within the larger system.

The effectiveness of the design principles is supported by research on fisheries, forestry, irrigation systems, and groundwater basins.[408]

407 Elinor Ostrom. 1990. Governing the Commons: The Evolution of Institutions for Collective Action. Cambridge University Press. Cambridge, United Kingdom.

408 Edella Schlager. 2004. Common pool resource theory, pp. 145–175 in *Environmental Governance Reconsidered: Challenges, Choices, and Opportunities* (Robert Durant, Daniel Fiorino, and Rosemary O'Leary, Eds.). MIT Press. Cambridge, Massachusetts.

Scientific, Local, and Traditional Knowledge

Local knowledge is knowledge obtained from embedded experiences. Traditional knowledge is a subset of local knowledge that describes indigenous and other forms of traditional knowledge about sustaining resources.

Both local and traditional knowledge typically include social, cultural, economic, and political aspects. Local and traditional knowledge is often required to augment sparse scientific and technical knowledge, reduce uncertainty, and to promote the management commitment of resource users.

The importance of local and traditional knowledge has long been recognized but challenging to link with scientific knowledge.[409] Local and traditional knowledge extends beyond science, can be qualitative rather than quantitative, use different terminology, have distinctive languages, nomenclatures, and taxonomies, and operate at a different scale. Challenges can be minimized by learning together with rather than learning about or from local and traditional partners.

Employing a problem-based, multiple evidence approach can dissolve the dichotomy between groups.[410] A multiple evidence approach emphasizes the complementary nature of various knowledge systems and avoids translating knowledge into one currency.[411]

The approach also recognizes the value of various natural and social science disciplines and recommends different validation criteria for each. Criteria include recognizing the spectrum of different knowledge holders, being open and respectful of diversity in views, cultures, and knowledge tools, committing to equitable benefit sharing, practicing reciprocity, establishing

409 Randy Thaman, Phil Lyver, Roger Mpande, Edgar Perez, Joji Carino, and Kazuhiko Takeuchi. (eds.) 2013. The Contribution of Indigenous and Local Knowledge Systems to IPBES: Building Synergies with Science. IPBES Expert Meeting Report, UNESCO/UNU. Paris, UNESCO.

410 Gemma Burford et al. 2012. Indigenous participation in intercultural education: Learning from Mexico and Tanzania. Ecology and Society 17(4):33. http://dx.doi.org/10.5751/ES-05250-170433.

411 Maria Tengo et al. 2013. The multiple evidence base approach as a framework for connecting diverse knowledge systems in IPBES. Discussion paper 2013-06-05. Stockholm Resilience Centre, Stockholm University, Sweden.

mutual trust and respect, and recognizing and respecting intellectual property.[412,413]

Scenario Planning

Scenario planning is a useful tool for improving understanding and decision-making in complex, uncertain situations common to socioecological systems. Scenarios are stories of what might be, narratives crafted to describe multiple plausible futures.[414,415,416] Scenarios stimulate collective learning and creativity, change mental models, and overcome biases associated with overconfidence and preference for confirming evidence.

The power of scenario planning is twofold. First, the interaction of participants of diverse knowledge, values, and experience facilitates individual and collective learning. Integrating scientific, local, and traditional knowledge.

Second, scenario planning is a platform for the collective action of scientists, policy makers, regulators, the public, and other stakeholders. This leads to thoughtful decisions and the stakeholder ownership needed for sustainability.

Scenarios are appropriate when there is a need to explore possibilities. Scenarios are forward-looking, focused on gathering information about what might happen based on the interests of participants, the probability of future events, and change drivers.

There are four fundamental types of scenarios.[417]

1. Vision scenarios examine the ideal future.

412 Randy Thaman, Phil Lyver, Roger Mpande, Edgar Perez, Joji Carino, and Kazuhiko Takeuchi. (eds.) 2013. The Contribution of Indigenous and Local Knowledge Systems to IPBES: Building Synergies with Science. IPBES

413 Rebecca Cross et al. 2017. Guidance for Integrating Indigenous and Local Knowledge in IUCN Red List Assessments. Final draft for Commission review/signoff.

414 Eva Wollenberg, David Edmunds, and Louise Buck. 2000. Anticipating change: Scenarios as a tool for adaptive forest management. Center for International Forestry Research. SMT Grafika Desa Putera, Indonesia. http://www.cifor.cgiar.org.

415 Kris Johnson. 2012. Using participatory scenarios to stimulate social learning for collaborative sustainable development. Ecology and Society 17(2):9. https://www.ecologyandsociety.org/vol17/iss2/art9/.

416 Paul Schoemaker. 1995. Scenario planning: A tool for strategic thinking. Winter. MIT Sloan Management Review. 36(2):25–40.

417 Eva Wollenberg, David Edmunds, and Louise Buck. 2000. Anticipating change: Scenarios as a tool for adaptive forest management. Center for International Forestry Research. SMT Grafika Desa Putera, Indonesia. http://www.cifor.cgiar.org.

2. Projection scenarios are conjectures about the future.
3. Pathway scenarios illustrate the path from the present to the future.
4. Alternative scenarios compare options through multiple vision, projection, or pathway scenarios.

A Structured Framework for Work in Socioecological Systems

Structured decision-making is another useful tool for working in socioecological systems. Structured decision-making is a general term for an organized analysis of a challenge for the purpose of finding a solution. Structured decision-making is informed by behavioral decision research and decision analysis.[418]

Behavioral decision research describes your cognitive shortcuts and poor ability to structure decision tasks, clarify objectives, and recognize influences when faced with a complex challenge. Decision analysis focuses on descriptive techniques to improve the quality of your decision-making. A structured decision-making approach helps you by dividing tough decisions into manageable parts.

The following is a structured decision-making process tailored to socioecological systems. In this example, conventional structured decision-making components - problem, objectives, alternatives, consequences, and trade-offs - are subsumed in a more organic process better suited to socioecological systems.

The structure below may vary depending on the circumstances and attributes of the system. Before beginning, decide whether your socioecological system is like a stakeholder meeting, peer-to-peer consult, a partners meeting, or hosting. The answer will guide the group's roles, responsibilities, and decision-making. Reaching an agreement will likely require multiple meetings over an extended period.

418 Robin Gregory, Lee Failing, Michael Harstone, Graham Long, Tim McDaniels, and Dan Ohlson. 2012. Structured Decision Making: A Practical Guide to Environmental Management Choices. Wiley-Blackwell. West Sussex, United Kingdom.

Issue Definition

The goals of this step are to gain a clear and common understanding of the issue and ensure all affected parties are represented. What is the issue you are trying to address? Does everyone agree that this is the issue? Why is the issue a problem that needs to be addressed? Does everyone agree with why this issue is a problem?

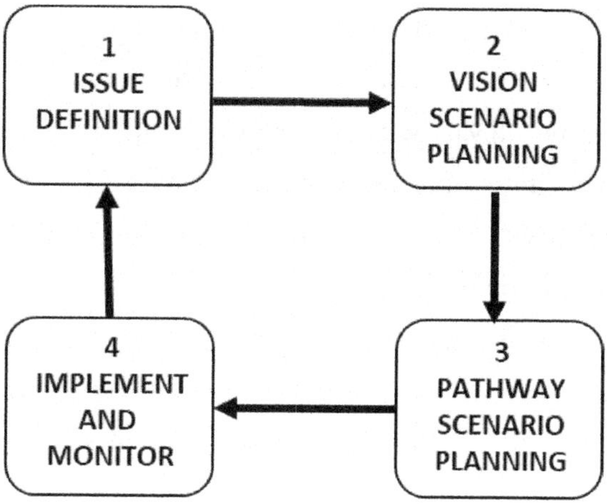

If the issue creates multiple problems, can they be prioritized? Is everyone represented that is affected by and needed to address the issue and the problems it creates? If not, who needs to be present?

Vision Scenario Planning

The purpose of vision scenario planning is to enable participants to articulate what their ideal future looks like and why it is important to them. The goal is to produce a single vision supported by the participants. Vision scenario planning replaces objective setting in a generic structured decision-making process.

The method works best when people can think freely about the desire for the future without being constrained by others' expectations or how to achieve their desired future. Ask participants what they would like to see changed and to independently craft a vision of desired future in as much detail as possible. Preferably this can be accomplished before the first meeting.

Elements of the vision might include value applied to each resource type, relative amount of each resource, resource condition, location of each resource type, type and extent of resource use, extent and location of modified lands and waters, and governance structure.

After each participant has crafted their vision, assemble them in groups of four or five. Each participant will share their vision with the others and engage in discussion. For large groups, you can employ the world café method of rotating participants among tables after a set period.

Talking points will vary with context but could include reasoning behind the vision, similarities and differences among visions, winners and losers for each vision, relative desirability of the visions, potential for merging the visions, and unresolved conflicts.

Harvest the learning of the participants after everyone has had an opportunity to share their vision with others. Gather the group and discuss key ideas, patterns, themes, knowledge gaps, and any questions that arose. Try and craft a single vision.

If you cannot craft a single vision, clarify objections, craft several visions, and ask each participant to rank the options. Then make the meeting material available to all participants and ask them to think about how they might craft a mutually acceptable vision of the future for discussion at the next meeting.

The time between vision scenario planning meetings is valuable for learning. For example, some participants may have little knowledge or experience with an ecological resource and would benefit from shadowing those with more knowledge and experience.

Similarly, the time could be used to become more sensitized to cultural norms. Opportunities for intimate learning build relationships that help people understand each other and increase their willingness to consider other perspectives and needs.

Pathway Scenario Planning

A pathway scenario helps participants determine how stakeholders can move from the present to the future imagined by their vision scenario. The goal is to reach agreement on a plan for achieving the collective vision. The approach focuses on open-ended problem-solving and creating strategies for overcoming constraints and realizing opportunities. Pathway scenario planning is a combination strategic and management plan.

Participants begin by characterizing the current resources, their condition, stakeholders, and stakeholder needs, beliefs, relationships, and levels of interest and authority. Working in groups of four or five individuals, participants contrast the present attributes with the desired future attributes to determine what must change.

Changes may be social and ecological. As with vision scenario planning, participants can rotate among groups and outcomes should be harvested to ensure agreement on what must change.

Knowledgeable about what must change, the participants re-assemble in small groups to discuss how to effect change. Are there multiple ways to achieve the vision? Are there constraints to change? Is so, can they be overcome or minimized?

How will resource practices change? How will cultural practices change? What new relations are needed? How will governance practices change? Who will provide labor and funding? What learning is required? What changes need to precede other changes? Facilitate discussion until everyone has had an opportunity to share thinking and then harvest the results.

Implement and Monitor

The pathway scenario plan is the basis for an implementation plan. Actions are defined, prioritized, and roles and responsibilities assigned. Activities, outputs, outcomes, and impacts can be articulated if capacity and capability are available. The socioecological system should be systematically monitored according to transparent monitoring indicators and evaluation criteria to determine progress toward the vision and to inform adjustments to the plan.

PART 7
HOPE FOR HEALTHY LANDS AND WATERS

HOPE FOR HEALTHY LANDS AND WATERS

We have always held to the hope, the belief, the conviction that there is a better life, a better world, beyond the horizon.

Franklin D. Roosevelt, 32nd president of the United States

Don't ever underestimate the importance you can have because history has shown us that courage can be contagious,
and hope can take on a life of its own.

Michelle Obama, attorney, administrator, writer and former first lady of the United States

The best way to not feel hopeless is to get up and do something. Don't wait for good things to happen to you. If you go out and make some good things happen, you will fill the world with hope, you will fill yourself with hope.

Barack Obama, 44th President of the United States

Science, sociology, and economics confirm that healthy lands and waters are at least as valuable as developed lands and likely more valuable. From the perspective of those who have less, healthy lands and waters are certainly more valuable.

We are all part of the same social, ecological, political, and economic system interacting with all living organisms, lands, and waters to form a synergistic, self-regulating, complex system that helps to maintain and perpetuate the conditions for life on earth.[419] Logic and self-interest argue that we should sustain ecosystem services and values if for no other reason than to preserve human well-being and prosperity.

We need to achieve a balance between developed and healthy lands and waters that sustains everyone. There is no us and them, only we, and the natural resources on which we depend. Collaboration and cooperation are fundamental to health, wealth, and happiness. Anything else is a short-term strategy with dire long-term consequences.

Yet, illogically healthy lands and waters continue to disappear and the value of remaining lands and waters continues to decline. We are losing ecosystem services, the value provided by healthy lands and waters, at the rate of US$4–20 trillion per year.[420]

The consequences are insupportable. Every year, more than 12 million people die prematurely due to environmental risks.[421] More than 750 million people lack access to clean drinking water. More than 800 million people do not have enough food to eat, and each year 9 million die of hunger.[422,423] We must do better.

Hope

Hope is the feeling that what is wanted can be achieved. A feeling of expectation and desire for a certain thing to happen.

Hope's strength is the strength of our desire. Hope is to want an outcome that makes your life better in some way. Hope can help make a tough situation more bearable.

419 James Lovelock and Lynn Margulis, L. (1974). Atmospheric homeostasis by and for the biosphere: the Gaia hypothesis. Tellus. Series A. Stockholm: International Meteorological Institute 26(1–2): 2–10.

420 Robert Costanza, Rudolf de Groot, Paul Sutton, Sander van der Ploeg, Sharolyn J. Anderson, Ida Kubiszewski, Stephen Farber, R. Kerry Turner. 2014. Changes in the global value of ecosystem services. Global Environmental Change 26:152–158.

421 World Health Organization. 2016. An estimated 12.6 million deaths each year are attributable to unhealthy environments. https://www.who.int/news-room/detail/15-03-2016-an-estimated-12-6-million-deaths-each-year-are-attributable-to-unhealthy-environments.

422 Do Something.Org. 2019. 3 September https://www.dosomething.org/us/facts/11-facts-about-global-poverty.

423 Mercy Corps. No date. Quick facts: What you need to know about global hunger. https://www.mercycorps.org/articles/quick-facts-what-you-need-know-about-global-hunger.

Envisioning a better future motivates us to take steps to make it happen. Everyone hopes for something so why not hope for healthy lands and waters? All else will follow.

If your goal is healthy lands and waters your story should inspire hope.[424] Hope motivates people to act. Hope assures people they can control the situation. Hope tells people that the actions they take are important and valued. The path to healthy lands and waters is paved with hope.

Public Opinion

In democratic countries, healthy lands and waters depend on public interest and support for conservation action. Hope is illustrated by the ever increasing public interest and support for environmental causes. In the United States alone, there are more than 44,000 environmental nonprofit organizations with cumulative revenues of more than $8 billion per year.[425] Forty six percent of United States environmental nonprofit revenue comes from grants and contributions. More than 8 million people are donors or members, and the number is rising.

Climate change is the most critical issue of our generation and for generations to come. The latest survey reveals that nearly 60 percent of people are alarmed or concerned by climate change.[426] For the first time more people are alarmed by climate change than concerned, cautious, disengaged, doubtful, or dismissive.

A 2019 Pew Research Center survey revealed that 67 percent of Americans think the federal government is not doing enough to reduce the effects of climate change. Seventy-seven percent believe developing alternative energy supplies is more important than increasing fossil fuel production.

424 Ronald Swainsgood and James Sheppard. 2010. The culture of conservation biologists: Show me the hope! BioScience 60(8):626–630.

425 Baird Straughan and Tom Pollak. 2008. The Broader Movement: Nonprofit Environmental and Conservation Organizations 1989–2005. The Urban Institute. Washington, D.C.

426 Matthew Goldberg, Abel Gustafson, Seth Rosenthal, John Kotcher, Edward Maiback, and Anthony. Leiserowitz. 2020. For the first time, the Alarmed are now the largest of Global Warming Six Americas. Yale University and George Mason University. New Haven, Connecticut. Yale Program on Climate Change Communication.

At least in the United States, public opinion favors healthy lands and waters. For hope to grow government and business leaders need to respect public opinion, review the science, and enact policies and practices that consider the long-term well-being and prosperity of people and economies.

Education

Carl Sagan, noted astronomer, cosmologist, astrophysicist, astrobiologist, author, science popularizer, and science communicator said, *If the citizens are educated and form their own opinions, then those in power work for us.*

The United States has too few citizens educated in science, technology, engineering, and math to make those in power work of us.[427] Our workforce is poorly equipped to solve the complex problem of healthy lands and waters. Too many citizens avoid forming their own opinion accepting without question the opinion of television, radio, and social media personalities conflicted by ideology and the pursuit of money.

Hope lies in shifts in our education system. Next Generation Science Standards, developed by the National Research Council, American Association for the Advancement of Science, Achieve, and 26 states, are now in 20 states and the District of Columbia. Twenty-four states have developed their own standards based on Next Generation Science Standards recommendations.

Paired with universal design for learning principles, Next Generation Science Standards ensure that all students are familiar with core science and engineering ideas, conduct science and engineering tasks, and can think like scientists and engineers when needed. Instruction provides multiple means for engagement, representation, action, and expression.

Hope lies in developing people's capacity to detect patterns and opportunities, to synthesize information and craft satisfying narratives, and to combine seemingly unrelated ideas into something new. Hope lies in creating people who can build an economy and a society built on inventive, empathic, big-picture learning.

427 National Science Teaching Association. Accessed 24 February 2020. https://ngss.nsta.org/why-standards-matter.aspx.

Hope comes from educating a generation of solutionaries.[428] This bold new approach sets as its goal a just, healthy, and humane world. The purpose of solutionary learning is to provide people with an engaging, personalized, real world oriented education that fully develops them in an interdependent world.

Tenets of solutionary learning include collaboration, multiple approaches, understanding connectedness, individual responsibility, discovery, developing solutions, and meaningful assessments of learning. The result is motivated, knowledgeable, and compassionate citizens able to solve real-world problems.

Policy

Hope for healthy lands and waters remains alive in the policy sector. There are signs of change though environmental policy admittedly tests our patience, tolerance, and compassion.

Nevertheless, policy makers persist in calls for transformative change - a fundamental, system-wide reorganization across paradigms, goals, and values called for by the Intergovernmental Science-Policy Platform on Biodiversity and Ecosystem Services[429]. And change that enables the globally connected society called for by the Millennium Assessment.[430]

Healthy lands and waters underlie each of the 17 United Nations Sustainable Development Goals supporting the 2030 Agenda.[431] The agenda is the world's growth strategy reflected in its interdependent goals and balance of environmental, social, and economic concerns.

The goals stimulate action over 15 years in areas of critical importance for humanity and the planet. National governments are responsible for delivering

428 Zoe Weil. 2016. The World Becomes What We Teach. Lantern Books. Brooklyn, New York.

429 Intergovernmental Science-Policy Platform on Biodiversity and Ecosystem Services. 2019. Global Assessment Report on Biodiversity and Ecosystem Services.

430 Millennium Ecosystem Assessment. 2005. Ecosystems and Human Well-being: Synthesis. Island Press, Washington, DC.

431 United Nations Environmental Programme. Accessed 24 February 2020. About the sustainable development goals. https://www.unenvironment.org/explore-topics/sustainable-development-goals/about-sustainable-development-goals#ourwork.

the agenda through policies, regulations, and incentives. The United Nations Environment Programme ensures implementation.

Hope is the World Bank, financer of large-scale development, updating its environmental and social framework after extended and meaningful collaboration with environmental and social experts.[432] The framework signifies the World Bank's commitment to sustainable development through policy and standards. A commitment to end extreme poverty and promote shared prosperity in a manner that benefits citizens and the environment.

Hope is the ongoing discussion about ecocide. In simple terms, ecocide is the destruction of the natural environment by deliberate or negligent human action. The initiative is committed to guaranteeing the livability of the planet by holding those responsible for ecosystem degradation and destruction legally and financially responsible. The proposed International Criminal Court law is supported by France, Belgium, the Netherlands, small island states, the Swedish workers movement, the Olof Palme International Centre, Greta Thunberg, and Pope Francis.

Hope is challenged in the United States. Despite the opinion and will of the majority, regressive, narrowly focused policies are sacrificing our future for short-term economic and political gains. We are experiencing an unprecedented unraveling of decades of thoughtful, legislated environmental safeguards.[433,434,435]

In hundreds of instances elected officials and industry leaders have turned their backs on climate change action, clean water, and protections for fish, wildlife, and lands and waters to the detriment of the public. Reinstituting a

432 The World Bank. 2016. World Bank Environmental and Social Framework. World Bank, Washington, D.C. https://www.worldbank.org/en/projects-operations/environmental-and-social-framework.

433 Sabin Centre for Climate Change Law. Climate Deregulation Tracker. https://climate.law.columbia.edu/climate-deregulation-tracker.

434 Harvard Environmental and Energy Law Program. Regulatory Rollback Tracker. https://eelp.law.harvard.edu/regulatory-rollback-tracker/.

435 Jesse Prentice-Dun. The Trump administration's unfinished business on public lands. 21 January. https://medium.com/west-wise/the-trump-administrations-unfinished-business-on-public-lands-5a8bda1c044e.

national commitment to healthy lands and waters requires patience and persistence.

Hope is alive at the regional and local levels in the United States. One example of many, the City of Miami, long vulnerable to hurricanes, storm surge, and flooding now exacerbated by climate change, created the Miami Forever Climate Ready plan.

A living document involving multiple departments and strategic partnerships, the plan focuses on heat, sea level rise, storms and impacts to resident well-being and prosperity.[436]. Solutions include increased tree canopy cover to cool ambient temperatures, parks, open spaces to mitigate flooding from sea level rise and storms, and protecting the city's coastline with natural and built structures to reduce erosion and flooding.

Science

There is hope for healthy lands and waters in increasingly innovative and cross-boundary science. The Nature Conservancy evolved from a land trust to innovators of distributional data for rare species and ecological communities. Then shifted to landscape-scale projects, and as currently constituted multi-objective planning that accounted for economic, resource, and community needs.[437]

Another environmental nonprofit, NatureServe, has long been a leader in assessing species and habitat vulnerability to climate change.[438,439] Recently, in partnership with ESRI, The Nature Conservancy, and Microsoft

436 City of Miami. 2020. Miami Forever Climate Ready. January. https://www.miamigov.com/Government/MiamiForeverClimateReady.

437 Peter Kareiva, Craig Groves, and Michelle Marvier. 2014. The evolving linkage between conservation science and practice at The Nature Conservancy. Journal of Applied Ecology. 51:1137–1147. https://besjournals.onlinelibrary.wiley.com/doi/epdf/10.1111/1365-2664.12259.

438 Bruce Young, Elizabeth Byers, Geoffrey Hammerson, Alan Redder, and Kristin Szabo. 2011. Guidelines for using the NatureServe Climate Change Vulnerability Index, version 2.1. https://www.natureserve.org/biodiversity-science/publications/guidelines-using-natureserve-climate-change-vulnerability-index.

439 Patrick Comer, Jon Hak, Marion Reid, Stephanie Auer, Keith Schulz, Healy Hamilton, Regan Smyth, and Matthew Kling. 2019. Habitat climate change vulnerability index applied to major vegetation types of the western interior United States. Land. 8: e108. https://www.natureserve.org/conservation-tools/climate-change-vulnerability-index-ecosystems-and-habitats.

NatureServe created the Map of Biodiversity Importance - an important tool for identifying the places most important for conserving at-risk species.[440]

Island Conservation gives hope for saving species from extinction. The organization has become one of the world's most effective environmental organizations by focusing on islands, innovatively removing plant and animal invasive species, and restoring native species. To date, Island Conservation has protected 487 species and subspecies from extinction on 64 islands[441] - a notable and unprecedented track record of achievement.

Perhaps the most hopeful science news for healthy lands and waters comes from Arctic Ice Project.[442] The organization is dedicated to climate restoration starting with restoring Arctic ice. Arctic ice plays a critical role in maintaining a safe and stable global climate. Arctic Ice Project has developed and tested a material that can restore ice through increasing reflectivity. If successful, the work will buy us time to reduce atmospheric greenhouse gases. Trillions of dollars in climate devastation, loss of life, and ecosystem destruction could be averted.

Socioecological Systems

I am most hopeful about the increasing development and impact of socioecological systems. Healthy lands and waters require more than regulation, education, policy, protected lands and waters, and science. Healthy lands and waters require motivated people working together to create processes, institutions, strategies, and plans. And then accepting responsibility for outcomes and impacts.

The signs are encouraging. Millions of people throughout the world are engaged in citizen science with the potential for many more. Scientists are at long last being exposed to community of practice techniques for engaging peers, partners, and a broad spectrum of stakeholders. Scientists are also

440 https://www.natureserve.org/conservation-tools/projects/map-biodiversity-importance.

441 Island Conservation. 2019. Impact Report. https://www.islandconservation.org/report/2019/wp-content/uploads/2019/11/Island-Conservation-Impact-Report-2019.pdf.

442 Leslie Field et al. 2018. Increasing Arctic sea ice albedo using localized reversible geoengineering. 23 June. Advancing Earth and Space Science. https://agupubs.onlinelibrary.wiley.com/doi/full/10.1029/2018EF000820.

recognizing the importance of local and traditional knowledge and partnering with nonscientists to manage common pool resources.

Abby Barrows, an independent research scientist, demonstrated the potential and power of citizen scientists in her global study of ocean microplastic and microfiber pollution.[443] Microparticles are ingested by marine life resulting in illness and death.

Abby and an army of citizen scientists collected water samples throughout the world's oceans over the course of five years producing the world's most extensive microparticle distribution dataset. The results documented pollution in all major ocean basins. Thanks to Abby and her citizen scientists we now have a better understanding of the movement and transformation of microparticles in regional seas and basins.

Hope occurs in the shape of International Rivers, an organization that protects rivers and defends the rights of communities that depend on them. International Rivers builds movements that empower citizens, ensures social and environmental issues are addressed, stops destructive projects, raises awareness about the human rights impacts of large dams, and promotes solutions. International Rivers has held up, prevented, or stopped 217 dam projects, created a network of 859 nonprofit partners around the world, and regranted money to 277 grassroots organizations.

Heroes arise in socioecological systems.[444] Chico Mendes was a rubber tapper in the Brazilian Amazon. Disturbed by government and rancher deforestation, Chico organized local rubber tappers and developed the concept of extractive reserves to protect the forest and their livelihoods. Extractive reserves use forests not only for rubber tapping but for gathering Brazil nuts, hunting for food, and growing subsistence crops in small patches.

Chico came to the attention of the Environmental Defense Fund and played a role in modifying World Bank and Inter-American Development Bank lending practices. For his sustainable system, Chico received the United

443 Abigail Barrows, Sara Cathey, and Chris Petersen. 2018. Marine environment microfiber contamination: Global patterns and the diversity of microparticle origins. Environmental Pollution. 237:275–284.

444 Larry Nielsen. 2017. Nature's Allies: Eight Conservationists Who Changed Our World. Island Press. Washington, D.C.

Nations Environment Programme's Global 500 Award in 1987. Tragically, he was assassinated in 1988 by those who opposed his movement.

In 2004, Wangari Maathai received the Noble Peace Prize for her contributions to sustainable development, democracy, and peace. Wangari lived and worked most of her life in Kenya, spending much of her time with rural women whose livelihoods were wracked by the conversion of native lands to exportable crops. Deforestation, devegetation, unsustainable agriculture, and soil loss were ruining Kenya.

Wangari's solution was to gather the women and plant native trees like fig, banana, citrus, papaya, avocado, and mango. Her Green Belt Movement planted more than 50 million trees in Kenya, changing the future for 900,000 Kenyan women and the country.

Needed: More Leaders

Leaders build hope in harnessing public opinion, educating, enacting policy, conducting science, and building socioecological systems. Leadership is scarce because too few people are willing to make the sacrifices required to lead.[445] Too many people want to be led, to believe in an idea, and to contribute to and take from a group of like-minded individuals. But the well-being and prosperity produced by healthy lands and waters requires that everyone, in their own way, become a leader.

The solution to healthy lands and waters can only be addressed by changing our thinking. We can only change the thinking of others by first changing ourselves. Leading people in the pursuit of healthy lands and waters requires the integrity, vision, competence, and humility to create a shared vision, build trust, and enable others.

Leadership requires understanding how your brain works so you can make good decisions, embrace change, work with others, de-escalate stress, and create and innovate. Leadership requires the patience, tolerance, compassion, and humility to help others learn from you and for you to learn from others.

445 Seth Godin: 2008. Tribes: We Need You to Lead Us.

The path to healthy lands and waters is predicated on people working together, having concerns for others, even when it does not directly benefit them. Humankind can, and will, collaborate.

Altruism can be promoted and has a physiological basis, triggering neural activity in areas of the brain responsible for reward processing.[446, 447] Despite the onslaught of negative news, collaboration, generosity, and altruism are not extinct. People will form collaborative networks for their own benefit.[448,449] And people will contribute to the greater good.

Aldous Huxley, writer and philosopher best known for the novel *Brave New World*, said, *It's a bit embarrassing to have been concerned with the human problem all one's life and to find at the end that one has no more to offer by way of advice than try to be a little kinder.*

If you are reading this you choose to lead, to help others understand the value of healthy lands and waters, and to act. To help others to lead, and so on. Aldous Huxley is wrong. Try to be a little kinder, it is a lot to offer.

446 William Harbaugh, Ulrich Mayr, and Daniel Burghart. 2007. Neural responses to taxation and voluntary giving reveal motives charitable donations. Science. Vol. 316, Issue 5831, pp 1622–1625.

447 Tom Farsides. 2007. The psychology of altruism. The Psychologist. Journal of the British Psychological Society (Digital). 20:474–477.

448 Orjan Bodin. 2017. Collaborative environmental governance: Achieving collection action in socioecological systems. Science. Vol. 357, Issue 659 (Digital).

449 Elinor Ostrom. 1990. Governing the Commons: The Evolution of Institutions for Collective Action. Cambridge University Press.

ACKNOWLEDGMENTS

Ideas arise not from the ether in an instant, but through openness to a lifetime of experiences no matter how bumpy. This idea – *Leadership Practices for Healthy Lands and Waters* – does indeed reflect a lifetime of experience, more than 30 years at last count. A lifetime of crossing discipline boundaries and challenging norms. Of straying from the path only to find a new path.

This idea has been with me throughout my life, reflecting a persistent attraction to natural lands and waters. For much of my life, the idea was fragmented and randomly held as a collection of memories and experiences, neglected while pursuing short-term obligations.

My early and long-standing participation in team sports likely impressed upon me the necessity of working with others. Supervisory assignments required a transition from being part of a team to leading a team, a recalibration and explicit awareness that a well-managed team is more productive and impactful than a poorly managed team or an individual.

My work at Walt Disney Imagineering Research and Development and on Florida Everglades restoration was transformative, forcing me to recognize that communication, education, and collaboration were essential skills if I were to convert scientific research into conservation action.

The final push was most explicit – design and implement a NatureServe leadership program that would help conservation scientists, many of whom were now for the first time, and without training, assuming management, policy, and fundraising responsibilities. My fragmented memories and experiences coalesced, revealing both the path I had taken, and the path ahead.

I benefited from an abundant leadership literature, beginning with the James Kouzes and Barry Posner's research-based work summarized in *The Leadership Challenge*. Others that influenced my thinking include Jim

Collins (*Good to Great*), Stephen M.R. Covey and Rebecca Merrill (*The Speed of Trust*), John Gookin and Sheri Leach (*National Outdoor Leadership School Educator Notebook*), Ronald Heifetz, Alexander Grashow and Marty Linsky (*The Practice of Adaptive Leadership*), Otto Scharmer (*Leading From the Future as it Emerges*) and Margaret Wheatley and Debbie Frieze (*Leadership in the Age of Complexity: From Hero to Host).*

I found inspiration in work not explicitly about leadership per se, but relevant to practices essential for leading. Sources include Rowan Gibson (*The 4 Lenses of Innovation*), Daniel Goleman (*Emotional Intelligence*), Jonah Lehrer (*How We Decide; Imagine, the Science of Creativity*), Kerry Patterson, Joseph Grenny, Ron McMillan, and Al Switzler (*Crucial Conversations: Tools for Talking When Stakes are High*), Daniel Pink (*A Whole New Mind; Drive)* and Simon Sinek (*Start with Why*).

Perhaps further afield, but just as important, were countless studies on brain science, psychology, creativity and innovation, strategy, business practices, sense of place, communication, and Buddhist philosophy. Inspiration is everywhere.

Most directly, I thank those I have had the good fortune to work with as a peer and instructor. Andrew Bennett of the Bennett Performance Group was instrumental in teaching me the importance of compassion and humanity to leadership practices. I had the privilege to work with the Conservation Leadership Programme's (Birdlife International, Wildlife Conservation Society, Flora and Fauna International, Conservation International) Christine Imrich, Robyn Dalzen, Maureen Ryan, and Christine Ageton, who collectively reinforced that leadership is not about me, but about everyone else.

My sincere gratitude to all the NatureServe staff and NatureServe Network individuals throughout the Western Hemisphere who trusted me with their learning, and provided valuable feedback on leadership program framework, content, implementation, and impact.

A special thanks to Andrew Bennett for a Foreword that sets the stage, and pre-release endorsements from Shane Ford, and Ricardo Stanoss.

Finally, but certainly not least, I thank my beloved wife Karen Thompson for her suggestions, support, and patience.

INDEX

ABOUT THE AUTHOR

Don is the principal and founder of Healthy Lands. For more than 30 years, Don has worked in socioecological systems to promote healthy, resilient, and sustainable lands and waters that support people and ecosystems.

Don has executed conservation initiatives in the United States and abroad with competencies in scientific research, strategy, planning, policy, management, communication, and leadership. He has 15 years of leadership experience in the nonprofit sector, and another 15 years in the private and government sectors.

Don is a Ph.D. in Ecology, Evolution, and Behavior, an AWAI-trained copywriter, and a member of the Professional Writers' Alliance. Career stops include an international consulting firm, a global entertainment company, state government, and a global biodiversity nonprofit.

Made in the USA
Monee, IL
04 November 2026

67387459R10167